# SpringerBriefs in Computer Science

More information about this series at http://www.springer.com/series/10028

Esther Galbrun • Pauli Miettinen

# Redescription Mining

 Springer

Esther Galbrun
Inria Nancy – Grand Est
Villers-lès-Nancy, France

Pauli Miettinen
Max-Planck-Institute for Informatics
Saarbrücken, Germany

ISSN 2191-5768          ISSN 2191-5776   (electronic)
SpringerBriefs in Computer Science
ISBN 978-3-319-72888-9          ISBN 978-3-319-72889-6   (eBook)
https://doi.org/10.1007/978-3-319-72889-6

Library of Congress Control Number: 2017963350

This Springer imprint is published by Springer Nature
The registered company is Springer International Publishing AG
The registered company address is: Gewerbestrasse 11, 6330 Cham, Switzerland

# Preface

'What is redescription mining?' is a question our colleagues and students often ask when we mention it as a topic we are working on. In short, redescription mining is the art of saying 'that is'. That is, redescription mining tries to describe the same phenomenon in two different ways. We usually augment this minimalistic definition with various details, leading to further questions such as 'Why do you use Jaccard similarity?', 'Can you apply your method to my data?', or 'What if you have more than two data sets?'

Wanting to provide answers for all of these questions, we prepared and presented tutorials on redescription mining. But not everybody could attend those tutorials, and although the slides are available online, they are not entirely self-explanatory. Moreover, there is a third group of persons who ask us those detailed questions and who are not satisfied if we say that we discussed the answers in our tutorial—the reviewers of our papers.

This short book is intended as an introduction to redescription mining, accessible to both practitioners and researchers alike. It develops a uniform framework in which the various formulations of redescription mining can be defined, explains the main algorithmic approaches used to mine redescriptions, and presents applications and variants of redescription mining, in addition to future research directions.

We hope this book will help those new to the field of redescription mining become familiar with the topic and let them consider how to apply redescription mining in their fields of interest. Those already familiar with redescription mining will hopefully find the book useful as a short reference work, and perhaps as a guide for further research directions. Reviewers for papers on redescription mining will hopefully find answers to some of their favourite questions in this book.

We are grateful to Krista Ames for providing language feedback for this book. Any mistakes that remain are—naturally—our own.

Saarbrücken, Germany
October 2017

Esther Galbrun
Pauli Miettinen

# Contents

# List of Figures

# List of Symbols

| Symbol | Description |
|---|---|
| $\mathcal{E}$ | Set of all entities in the data |
| $\mathcal{A}$ | Set of attributes of the entities |
| $\mathcal{V}$ | Set of views the attributes can be divided into |
| **D** | Entities-by-attributes table corresponding to one view |
| $\mathcal{D}$ | The data |
| $\mathcal{P}$ | Set of all predicates over $\mathcal{E} \times \mathcal{A}$ |
| $\mathcal{L}$ | Set of all literals, i.e. all predicates and their negations |
| $\mathcal{Q}$ | Query language, i.e. a set of Boolean functions over $\mathcal{L}$ |
| supp($q$) | Support (set) of a query $q$ |
| att($q$) | Attributes of a query $q$ |
| views($q$) | Views of a query $q$ |
| $d(p, q)$ | Distance between the supports of queries $p$ and $q$ |
| $p \sim q$ | Set supp($p$) is similar to set supp($q$) |
| $p \equiv q$ | Sets supp($p$) and supp($q$) are the same |
| $J(p, q)$ | Jaccard similarity between supp($p$) and supp($q$) |

# Chapter 1
# What Is Redescription Mining

What is redescription mining? The answer to the eponymous question of this chapter involves some amount of theoretical framework-building: definitions that are used to make other definitions that in turn are used to define yet new concepts that—hopefully—finally yield a coherent and complete definition of redescription mining. That, at least, is the mathematical way to answer the question. A more holistic approach would be to consider how redescription mining relates to other data analysis methods, defining it not by what it is, but through its similarities and dissimilarities. Or, perhaps one could define redescription mining by looking at its evolution, asking how it started and how it became what it is.

These are three valid approaches for defining redescription mining, and we will examine them in this chapter. First, though, let us answer the titular question of this chapter with an ostensive definition of redescription mining.

## 1.1 First Examples of Redescriptions

Consider an ecologist who wants to understand what kind of bioclimatic environment different mammal species require. She knows the regions the different mammal species inhabit, and she knows the bioclimatic conditions of those places, such as monthly average temperatures and precipitation. In ecology, her task is known as *bioclimatic niche (or envelope) finding* (Soberón and Nakamura 2009).

What would such a niche[1] look like? One example could say that:

> The areas inhabited by either the Eurasian lynx or the Canada lynx are approximately the same areas as those where the maximum March temperature ranges from $-24.4\,°C$ to $3.4\,°C$.

---

[1] We use the term *niche* as defined by Grinnell (1917).

© The Author(s) 2017
E. Galbrun, P. Miettinen, *Redescription Mining*, SpringerBriefs
in Computer Science, https://doi.org/10.1007/978-3-319-72889-6_1

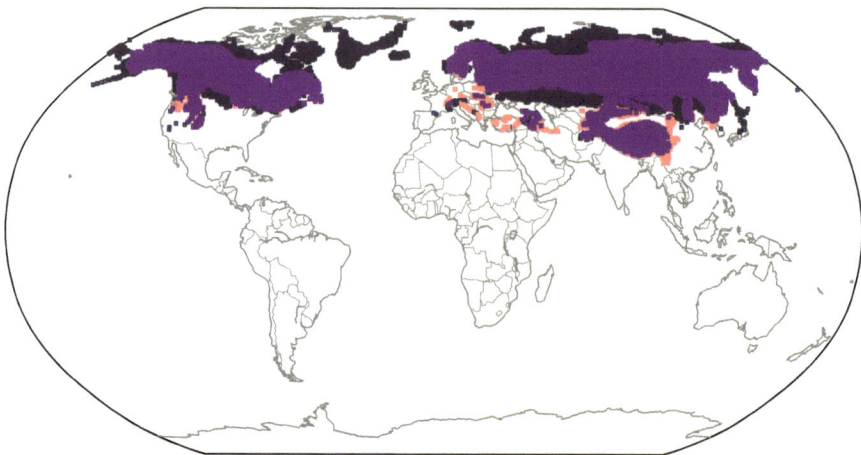

**Fig. 1.1** The areas inhabited by either the Eurasian lynx or the Canada lynx (light red and medium purple) and the areas where the maximum March temperature is between $-24.4\,°C$ and $3.4\,°C$ (dark blue and medium purple)

The above sentence *describes* areas of the earth in two different ways; on the one hand, by the fact that certain species inhabit them, and on the other hand, by the fact that they have a certain climate. We can see the areas described above in Fig. 1.1. Visualizations of redescriptions help us to interpret the redescriptions and understand what they describe and how. We will explain different visualization techniques throughout the book as we encounter them. These explanations are set in a distinctive block, as below.

**Visualization: Maps**
Maps are an easy way to visualize redescriptions over geographical regions. Everybody is familiar with maps and knows how to read them. Their use, naturally, requires that the data is associated with geographical locations. In Fig. 1.1, the medium purple colour denotes the areas where both of the above conditions hold (inhabited by one of the lynx species and with maximum March temperatures in the correct range), light red denotes the areas inhabited by one of the lynx species but where March temperatures are out of the range, and dark blue denotes the areas where the maximum March temperature is in the correct range but neither of the lynxes is found.

The above explanation of a niche is an example of a *redescription*: it provides two different ways to describe the (approximately) same regions on the earth. In short, a redescription is simply a pair of descriptions, both describing roughly the

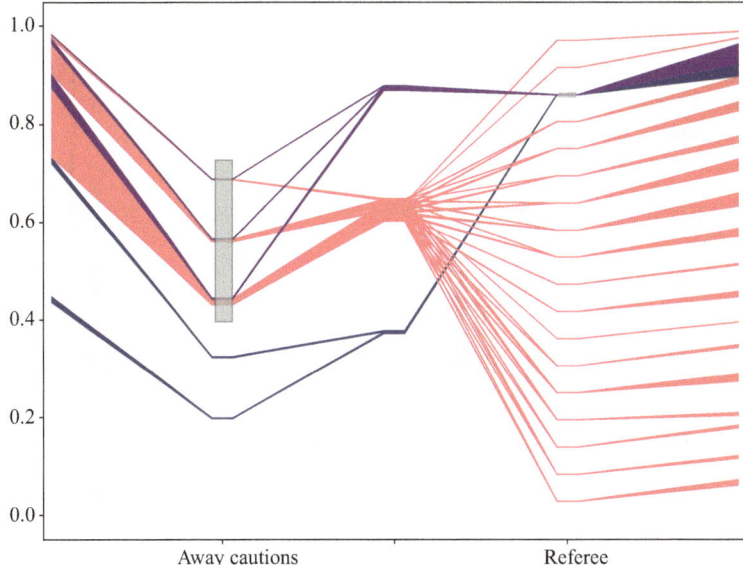

**Fig. 1.2** Parallel coordinates plot for the matches in the Premier League's 2013/14 season. The matches where the away team players were cautioned three to five times are in light red or medium purple, and the matches where Phil Dowd was the referee are in dark blue or medium purple. Very light grey lines correspond to matches where neither of the conditions hold

same entities (here, geographical regions). And, as we can see from this example, both the descriptions and what they describe can be of interest. The ecologist is interested in the descriptions in order to understand the *model* of the niche and in the geographical areas in order to understand where the niche holds (or does not hold).

Let us consider another example application of redescriptions. This time, an avid association football[2] fan wants to analyse the statistics of the matches in his favourite game. Specifically, he analyses the matches played in the UK's Premier League in the 2013/14 season. What he finds is the following:

> In two thirds of the matches where Phil Dowd was the referee, the away team players were cautioned three to five times.

We can see this result visualized in Fig. 1.2 using a *parallel coordinates* plot (Inselberg 2009).

---

[2]The game is known as either *soccer* or *football*; we will use the latter term, but want to emphasize that the game must not be confused with gridiron football, rugby football, Australian rules football, or Gaelic football.

**Visualization: Parallel Coordinates**

Parallel coordinates are an intuitive way to visualize entities with many attributes. They are also well-suited for visualizing redescriptions from the point of view of the descriptions rather than of the entities being described. Parallel coordinates plots can be used to visualize any redescription, though if the descriptions are complex, the visualization gets harder to interpret.

To understand how parallel coordinates visualize a redescription, consider Fig. 1.2. Here, each match from Premier League's 2013/14 season is represented by a line running from the left to the right of the plot. The lines are depicted by the same colours as the geographical regions in Fig. 1.1: medium purple corresponds to matches where the away team players were cautioned three to five times and Phil Dowd was the referee, light red lines represent the matches where the away team players were shown the yellow card three to five times but the referee was somebody other than Mr. Dowd, and dark blue lines represent the games where Mr. Dowd was the referee, but he cautioned the away team players fewer than three times. (In no matches did he show the yellow card to the away team players more than five times.)

Figure 1.2 also tells us about the match statistics related to the features used in the above descriptions. There are three virtual vertical axes in the plot, one in the middle, and two labelled as *Away cautions* and *Referee*. The point where a line crosses these axes indicates the value of that feature in the corresponding match; for example, a line that crosses the grey box in the *Away cautions* axis corresponds to a match where the away team players were shown the yellow card three to five times, and a line that crosses the (small) grey box in the *Referee* axis corresponds to a match where Phil Dowd was the referee. All values are scaled to unit range to facilitate the drawing.

From Fig. 1.2, we can see that there were many matches where the away team players were cautioned three to five times by a referee other than Phil Dowd, but he showed the yellow card to the away team players three to five times in all but ten of the 26 matches he refereed that season.

The above examples give some understanding of what redescriptions are and what a redescription looks like. While such redescriptions could be constructed manually, the goal of redescription mining is to find them automatically without any information other than the raw data (and some user-provided constraints). In the bioclimatic niche finding example, the ecologist should not have to define the species she is interested in, and in the football example, the analyst should not have to define his interest in yellow cards or referees. Rather, the goal of redescription mining is to find all redescriptions that characterize sufficiently similar sets of entities and adhere to some simple constraints regarding, for example, their type and complexity and how many entities they cover.

**Definition 1 (Redescription Mining, Informal Definition)** A *redescription* is a way to characterize roughly the same objects in two (or more) different ways. The goal of *redescription mining* is to find all redescriptions that satisfy specified constraints.

In the following, we will formalize this informal definition.

## 1.2 Formal Definitions

The formalization of redescription mining we present here comes mostly from Galbrun (2013). In addition to the general formalization, we will also present a somewhat simplified version that is suitable for most applications, though its simplicity makes certain definitions and extensions more cumbersome than necessary.

### 1.2.1 The Data

The basic building block of the redescription data model is the set of *entities* $\mathcal{E}$. Each entity $e \in \mathcal{E}$ is associated with a set of *attributes* from $\mathcal{A}$. The value of attribute $a$ in entity $e$ is denoted by $a(e)$. To keep the notation clean, we write $a$ to denote both the actual attribute and its value in some unspecified entity. Entities and their attributes comprise the data.

Redescriptions aim at providing different views on the same data. We assume that the attributes are partitioned into *views* $\mathcal{V} = \{V_1, V_2, \ldots, V_k\}$. $\mathcal{V}$ is a partition of $\mathcal{A}$, that is, $V_i \cap V_j = \emptyset$ for all $i \neq j$ and $\bigcup_{i=1}^{k} V_i = \mathcal{A}$. The view to which attribute $a$ is assigned is denoted by $\text{view}(a)$.

The set of attributes $\mathcal{A}$ must always be divided into at least two views; if the data does not have any natural partition, we can use the finest partition of $\mathcal{A}$: there are $|\mathcal{A}|$ views, and each view $V_i$ is a singleton containing attribute $a_i \in \mathcal{A}$. The reason for this restriction will become clear later.

The formal definition of the data model is:

**Definition 2 (Data Model)** The *data* $\mathcal{D}$ for a redescription problem is a tuple $\mathcal{D} = (\mathcal{E}, \mathcal{A}, \mathcal{V})$, where *entities* $e \in \mathcal{E}$ are associated with *attributes* from $\mathcal{A}$, and the attributes are partitioned in *views* $V \in \mathcal{V}$.

*Example 1* In the ecological niche finding example above, the entities are the geospatial regions, and their attributes are the species inhabiting those regions and climate variables. The attributes are divided into two views: the species and the climate variables. In the football example, the entities are the matches, while the attributes are the various match statistics, such as the number of goals, shots, corners, free kicks, etc. There is no natural division of these attributes, and hence, each view is a singleton set containing just one attribute.

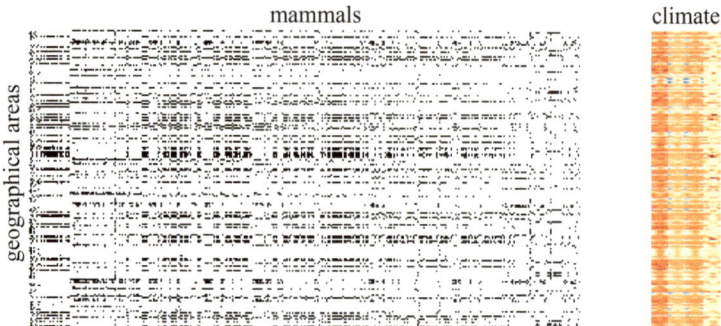

**Fig. 1.3** Example of the mammals-and-climate data as two tables. The table on the left has Boolean values and indicates which mammals (columns) are observed in which regions (rows). The table on the right has numerical values and contains the climate variables (columns) over the same regions (rows) as in the table on the left

We will need the above general framework when we discuss variants of the redescription mining problem in Chap. 3. For most applications of redescription mining, the simpler data model below is perfectly adequate.

**Definition 3 (Table-Based Data Model)** In the *table-based data model*, the data consist of one or more *tables* $\mathbf{D}_1, \mathbf{D}_2, \ldots$. The rows of the tables are the entities, and each table's rows must correspond to the same entities. The columns of the tables are the attributes, and the value $\mathbf{D}_k(i, j)$ is the value that entity $i$ takes for attribute $j$ in table $k$.

In this simplified model, the views are implicitly encoded in the tables: each view corresponds to one table, except in the case when all views are singletons; in this case, there is only one table.

*Example 2* The data the ecologist would use in the bioclimatic niche finding application has a natural interpretation as a pair of tables. The first table contains the binary attributes describing the presence of mammal species in different geographical regions, while the second table describes the climate of those regions. A subset of such data is presented in Fig. 1.3.

## 1.2.2   The Descriptions

Our definition of attributes leaves open their types. Indeed, as we saw in the examples, the attributes can be numerical (e.g. temperature or number of yellow cards), categorical (e.g. the referee of the match), or Boolean (e.g. whether a species inhabits a certain region or not), and might not even be observed for every entity. To form the descriptions, we need to endow entity–attribute pairs with *predicates*.

A predicate for attribute $a \in \mathcal{A}$ is a function $p_a \colon \mathcal{E} \to \{\text{true}, \text{false}\}$ that assigns a truth value for this attribute in every entity. The set of *literals* $\mathcal{L}$ contains all predicates and their negations.

What the predicates look like depends on the type of the attributes and on the user's needs (or algorithm's capabilities). We write a predicate for attribute $a$ of entity $e$ as $p_a(e) = [P_a(e)]$, where $P_a(e)$ is some logical proposition involving the value of attribute $a$ in entity $e$. The predicate is true if the proposition is true, and false otherwise. For Boolean attributes, their predicates simply return the value of the attribute in the entity. For categorical attributes, the propositions can be more complicated. In this book, we mostly consider the equivalence proposition, $p_a(e) = [a(e) = X]$, where $X$ is some constant. We write this as $[a = X]$ when we mean the predicate as a function. Another possible proposition for categorical attributes could be the set inclusion proposition, $[a \in \{X, Y, Z, \ldots\}]$.

For numerical attributes, we mostly concentrate on propositions that test whether the attribute's value is in some interval or half-line, that is, on predicates $[x \le a \le y]$, $[a \le x]$, and $[a \ge x]$. Much more complex predicates could be built by applying arbitrary functions on the values of the numerical attributes; for instance, we could consider predicates of the type $[\sqrt{a} \le x]$ that would test whether the square root of the value of attribute $a$ is below some threshold $x$. We do not consider such predicates here, though. First, for many functions, the equivalent result can be obtained by transforming the threshold (e.g. $[a \le x^2]$), and as the predicates are built by the redescription mining algorithms, adding the functions in the predicates would needlessly complicate them. Second, for the functions where we cannot easily move the transformation to the threshold (e.g. $[\sin(a) \le 0]$), we can obtain the desired result by adding a new attribute containing the transformed value.

A *description* is a Boolean query $q \colon \mathcal{E} \to \{\text{true}, \text{false}\}$ over the literals that assigns a truth value to each entity $e \in \mathcal{E}$. The query can, in principle, be an arbitrary Boolean function of the literals, such as $(\ell_1 \wedge \ell_2) \vee \neg (\ell_1 \vee \ell_3) \wedge (\ell_2 \vee \ell_3)$, but such queries are often difficult to interpret and have a great risk of overfitting, in addition to being difficult to optimize. Hence, it is common to restrict the queries to some *query language* $\mathcal{Q}$. Examples of query languages include *monotone conjunctive queries*, *linearly parsable queries*, and *tree-shaped queries*. Monotone conjunctive queries (e.g. $a \wedge b \wedge c \wedge d$) are the simplest of queries, and they are also often the easiest to work with. Algorithms based on itemset mining are often limited to such queries (see Sect. 2.1). Linearly parsable queries (e.g. $((a \vee b) \wedge \neg c) \vee d$) draw their names from the fact that their parse tree looks like a line: all the operations are evaluated from left to right, ignoring the normal operator precedence, no variable can appear more than once, and negations are only applied to predicates. Hence, linearly parsable queries can be written without parentheses with the convention that all operators have the same precedence; the query above, for example, would turn into $a \vee b \wedge \neg c \vee d$. This structure is amenable for algorithms building the queries literal-by-literal (see Sect. 2.3). Tree-shaped queries (e.g. $(a \wedge b) \vee (\neg a \wedge c) \vee (\neg a \wedge \neg b \wedge c)$) encode a decision tree (the clauses are different paths down the tree) and have a very particular, and sometimes unintuitive, form. They are most commonly seen with methods that are based on decision tree induction (see Sect. 2.2).

*Example 3* The descriptions we saw at the beginning of this chapter can now be expressed more formally. The query corresponding to *'The areas inhabited by either Eurasian lynx or Canada lynx'* could be written as

$$Eurasian\ lynx \lor Canada\ lynx\ ,$$

where *Eurasian lynx*$(e)$ = true if Eurasian lynx inhabits the area $e$. The description *'maximum March temperature ranges from* $-24.4\,°C$ *to* $3.4\,°C$*'* could be written as

$$[-24.4 \le t_3^+ \le 3.4]\ ,$$

where $t_3^+$ is the attribute for the maximum March temperature. The descriptions in the football example would be

$$[3 \le away\ cautions \le 5] \quad \text{and} \quad [referee = P.\ Dowd]\ .$$

The queries can be associated with three important sets: support, attributes, and views of a query.

**Definition 4 (Support, Attributes, Views)** Let $q \in \mathcal{Q}$ be a query. Its *support*, supp$(q)$, is the set of entities that evaluate true in $q$, that is, supp$(q) = \{e \in \mathcal{E} : q(e) = \text{true}\}$. The *attributes* of $q$, att$(q)$, is the set of attributes that appear in $q$. The *views* of $q$, views$(q)$, is the union of all views of all attributes in att$(q)$: views$(q) = \bigcup_{a \in att(q)}$ view$(a)$.

Notice that many authors in data analysis use the term *support* to refer to the size (or cardinality) of what we call support. That is, what they call support of $q$ is the size of the support, $|\text{supp}(q)|$, for us. Conversely, what we call support is sometimes referred to as the *support set*.

## 1.2.3   The Redescriptions

A *redescription* is a pair $(p, q)$ of descriptions with disjoint views and with sufficiently similar supports. The second constraint is easy to understand. If the queries' supports are very different, they do not 'explain (approximately) the same entities'. To formalize this, we need some way to measure the difference between two sets of entities. To that end, we can use any distance function $d: 2^{\mathcal{E}} \times 2^{\mathcal{E}} \to [0, \infty)$. We require that $d$ is at least a semimetric, that is,

$$d(X, Y) = 0 \qquad\qquad \text{if and only if } X = Y; \text{ and} \qquad (1.1)$$

$$d(X, Y) = d(Y, X) \qquad\qquad \text{for all } X, Y \in 2^{\mathcal{E}}\ , \qquad (1.2)$$

although it is often a metric with a triangle inequality

$$d(X, Y) \le d(X, Z) + d(Z, Y) \qquad \text{for all } X, Y, Z \in 2^{\mathcal{E}}\ . \qquad (1.3)$$

For a pair of queries $(p, q)$, we write $d(p, q)$ to denote $d(\text{supp}(p), \text{supp}(q))$. The most common choice for a distance measure between descriptions is the *Jaccard distance*, which is based on the *Jaccard similarity index*:

**Definition 5** The *Jaccard (similarity) index $J$* between the supports of two descriptions $p$ and $q$ is defined as

$$J(p, q) = J(\text{supp}(p), \text{supp}(q)) = \frac{|\text{supp}(p) \cap \text{supp}(q)|}{|\text{supp}(p) \cup \text{supp}(q)|} . \tag{1.4}$$

The *Jaccard distance* is defined as

$$1 - J(p, q) = 1 - \frac{|\text{supp}(p) \cap \text{supp}(q)|}{|\text{supp}(p) \cup \text{supp}(q)|} . \tag{1.5}$$

Definition 5 requires that either $\text{supp}(p) \neq \emptyset$ or $\text{supp}(q) \neq \emptyset$. As the descriptions are supposed to describe something, we require that supports of both $p$ and $q$ are non-empty. The Jaccard distance is not the only possible distance function, but since it is the most common one, we will postpone the discussion of other alternatives until Sect. 1.2.5.

With a distance function, we can measure how similar the supports of two queries are. If the supports are the same, that is, $d(p, q) = 0$, we say that the redescription $(p, q)$ is *exact* and write $p \equiv q$.

Most redescriptions are not exact, however, and we are often content with descriptions that are similar enough. But how similar is 'similar enough'? That is something the user must decide, depending on the data, her needs, and the selected distance function. Therefore, we say that the supports of $p$ and $q$ are similar enough if $d(p, q) \leq \tau$ for some user-specified constant $\tau \in [0, \infty)$. We denote this by $\sim_\tau$, that is,

$$p \sim_\tau q \quad \text{if and only if} \quad d(p, q) \leq \tau \tag{1.6}$$

and drop the subscript $\tau$ in most cases, writing just $p \sim q$. Notice that $\sim$ is not transitive, that is, it is possible that $p \sim q$ and $q \sim r$ but $p \not\sim r$.

The distance—in addition to measuring whether the descriptions are similar enough—is also often used to measure the *quality* of the redescription. The smaller the distance between the two descriptions, the better the redescription (provided it adheres to the other constraints, naturally).

The first constraint for a pair $(p, q)$ to be a valid redescription is that their views must be disjoint, that is, $\text{views}(p) \cap \text{views}(q) = \emptyset$. Why such a constraint? The goal of a redescription is to provide *different* ways to describe the same entities. At minimum, this requires that the descriptions have different attributes because, for example, a redescription that says '*the regions inhabited by the polar bear are the regions not inhabited by the kangaroo and inhabited by the polar bear*' would not be very interesting at all. This constraint could be achieved by just requiring that the attributes are disjoint, that is, $\text{att}(p) \cap \text{att}(q) = \emptyset$. When each attribute is mapped to a singleton view, this constraint becomes equivalent with the constraint that the views of the queries are disjoint.

But many data sets have natural divisions of the attributes; for example, in the bioclimatic niche finding example, we can naturally divide the attributes into the mammal species' inhabitancy attributes and climate attributes. With such data sets, we often want to restrict the queries to take attributes only from one view (e.g. one query over the inhabitancy attributes and another over the climate attributes). This can also be achieved by dividing the attributes into two views and requiring that the views of the queries are disjoint.

We can now define what a redescription is.

**Definition 6 (Redescription)** A *redescription* in query language $\mathcal{Q}$ over data $\mathcal{D}$ with similarity $\sim$ is a pair of queries $(p, q) \in \mathcal{Q} \times \mathcal{Q}$ such that

$$p \sim q \quad \text{and} \quad \text{views}(p) \cap \text{views}(q) = \emptyset . \tag{1.7}$$

*Example 4* We can now write the redescriptions from Sect. 1.1 (using the queries from Example 3) as follows:

$$\textit{Eurasian lynx} \vee \textit{Canada lynx} \sim [-24.4 \le t_3^+ \le 3.4]$$
$$[3 \le \textit{away cautions} \le 5] \sim [\textit{referee} = P.\, Dowd] .$$

Defining the redescriptions in the simplified table-based data model of Definition 3 is in fact a bit more cumbersome. The similarity function stays the same, but the requirement of disjoint views changes depending on how many tables are used.

**Definition 7 (Redescription in the Table Data Model)** If the data consist of one table $\mathbf{D}$, a redescription over that table is a pair of queries $(p, q) \in \mathcal{Q} \times \mathcal{Q}$ such that $p \sim q$ and $\text{att}(p) \cap \text{att}(q) = \emptyset$. If the data consist of two tables, $\mathbf{D}_1$ and $\mathbf{D}_2$, we instead require that $\text{att}(p) \subseteq \text{att}(\mathbf{D}_1)$ and $\text{att}(q) \subseteq \text{att}(\mathbf{D}_2)$, where $\text{att}(\mathbf{D})$ denotes the set of attributes that correspond to the columns of $\mathbf{D}$.

The definition of a support of a description extends naturally to redescriptions.

**Definition 8 (Support of a Redescription)** Let $(p, q)$ be a redescription in data $\mathcal{D}$. The *support* of $(p, q)$ is the support of query $p \wedge q$ in $\mathcal{D}$, that is,

$$\text{supp}(p, q) = \text{supp}(p \wedge q) = \text{supp}(p) \cap \text{supp}(q) .$$

Consider a pair of queries $(p, q)$. The supports of queries $p$ and $q$ and the support of the redescription $(p, q)$ define four important sets of entities:

$$\mathcal{E}_{11} = \text{supp}(p) \cap \text{supp}(q) , \tag{1.8a}$$
$$\mathcal{E}_{10} = \text{supp}(p) \setminus \text{supp}(q) , \tag{1.8b}$$

**Fig. 1.4** A Venn diagram
showing the relationships
between the sets in (1.8)

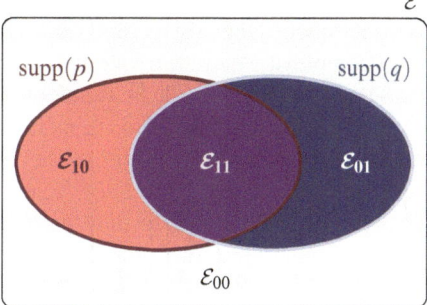

$$\mathcal{E}_{01} = \text{supp}(q) \setminus \text{supp}(p) , \text{ and} \tag{1.8c}$$

$$\mathcal{E}_{00} = \mathcal{E} \setminus \big(\text{supp}(p) \cup \text{supp}(q)\big) . \tag{1.8d}$$

Figure 1.4 illustrates the relationships between these sets and how they correspond to the colours used in the visualizations.

The goal of redescription mining is to find all valid redescriptions from the data.

**Definition 9 (Redescription Mining)** Given data $\mathcal{D}$, query language $\mathcal{Q}$, similarity $\sim$, and other potential constraints, the goal of *redescription mining* is to find all valid redescriptions $(p_i, q_i)$ that also satisfy the other potential constraints.

## 1.2.4 Other Constraints

In the above definition of redescription mining, the redescriptions were restricted to those that 'also satisfy the other potential constraints'. Why have other constraints, and what could these be? The main purpose of the constraints is to remove unwanted redescriptions. What kind of redescriptions are unwanted depends strongly on the application and on the data.

Some of the most common constraints limit the total support of the redescription. The support can be bounded either from above or from below. Rejecting redescriptions that have too small a support is intuitive; such redescriptions do not describe many entities, and hence, do not (usually) provide interesting insights on the data. Redescriptions with too large a support, on the other hand, are often uninteresting because they either describe a general tautology in the data, or—especially if the query language allows negations and disjunctions—they cover the entities using negations of redescriptions with very small a support, or by chaining unrelated attributes with disjunctions. For example, a redescription on football that says

> those games where the home team made at least zero goals are exactly those games where the away team made at least zero goals

is not very interesting, as neither team can make *fewer than* zero goals, and hence, the redescription covers every football match ever played.

The *complexity* of the redescriptions is another common constraint. Choosing the correct query language is one—and arguably the most powerful—way to control the complexity of redescriptions, but another way is to limit the *length* of the descriptions. The length of the descriptions is usually defined as the number of literals that appear in them, though in the case of tree-shaped descriptions (see Sects. 1.2.2 and 2.2), the depth of the tree is potentially a more intuitive measure.

Measuring the *statistical significance* of the found redescriptions is yet another way of removing uninteresting results. One can think of different null hypotheses to identify uninteresting redescriptions, but most of the existing literature uses the simple null hypothesis proposed originally by Ramakrishnan et al. (2004). The null hypothesis is that the supports for $p$ and $q$ are random independent sets with expected sizes of $|\mathrm{supp}(p)|$ and $|\mathrm{supp}(q)|$, respectively, and the associated $p$-value is the probability that two such sets overlap as much as they do. This probability is a tail of the binomial distribution. Let $X \subseteq \mathcal{E}$ and $Y \subseteq \mathcal{E}$ be two random independent sets such that $\Pr(e \in X) = |\mathrm{supp}(p)|/|\mathcal{E}|$ and $\Pr(e \in Y) = |\mathrm{supp}(q)|/|\mathcal{E}|$ for all $e \in \mathcal{E}$, and let $\alpha$ be the probability that some random $e \in \mathcal{E}$ is in $X \cap Y$. Denoting $|\mathcal{E}|$ by $n$ and using the independency of $X$ and $Y$, we get that

$$
\alpha = \Pr(e \in X, e \in Y) = \Pr(e \in X)\Pr(e \in Y) = \frac{|X|}{n}\frac{|Y|}{n} = \frac{|\mathrm{supp}(p)||\mathrm{supp}(q)|}{n^2} .
$$

(1.9)

The size of $X \cap Y$ is binomially distributed with probability $\alpha$ and maximum size $n = |\mathcal{E}|$, and hence, the probability that $|X \cap Y| \geq |\mathrm{supp}(p,q)|$ is

$$
\sum_{k=|\mathrm{supp}(p,q)|}^{n} \binom{n}{k} \alpha^k (1-\alpha)^{n-k} ,
$$

(1.10)

which is the desired $p$-value.

This null hypothesis favours redescriptions where the supports of the queries are small; the probability that two independent small sets have a large overlap is quite small, while if the sets are large (and hence $\alpha$ is large), they are expected to have a large overlap simply by chance. The $p$-value can be used as a constraint for redescriptions in a natural way: the user can define the maximum $p$-value that still corresponds to significant redescriptions (e.g. 0.01 or 0.05), and all redescriptions with higher a $p$-value can be removed.

The above $p$-value calculation fixes the *expected* sizes of the supports for queries $p$ and $q$. Alternatively, we can also fix the sizes to the true sizes and calculate the $p$-value based on fixed support sizes. This is equivalent to the one-sided $p$-value in Fischer's exact test and can be calculated using the hypergeometric distribution. Using the same notation as above, we get

$$
\sum_{k-|\mathrm{supp}(p,q)|}^{n} \frac{\binom{|\mathrm{supp}(p)|}{k}\binom{|\mathcal{E}|-|\mathrm{supp}(p)|}{|\mathrm{supp}(p)|-k}}{\binom{|\mathcal{E}|}{|\mathrm{supp}(q)|}} .
$$

(1.11)

Another way to measure the significance is to consider the queries themselves. Consider first a query $q$ with just one attribute, $a$. Assuming that the attributes' values in the entities are independent, the probability of the query having the support it has is $\alpha_q = \text{supp}(q)/|\mathcal{E}|$. For more complex queries $q$, the probability under the independence assumption can be defined recursively:

$$
\alpha_q = \begin{cases} \alpha_{q_1}\alpha_{q_2} & \text{if } q = q_1 \wedge q_2 \\ 1 - \alpha_{q_1} & \text{if } q = \neg q_1 \\ \alpha_{q_1} + \alpha_{q_2} - \alpha_{q_1}\alpha_{q_2} & \text{if } q = q_1 \vee q_2 \, . \end{cases} \tag{1.12}
$$

Similarly to (1.10), the total probability of seeing a query $q$ with support $\text{supp}(q)$ or higher is a tail of the binomial distribution:

$$
\sum_{k=|\text{supp}(q)|}^{n} \binom{n}{k} \alpha_q^k (1 - \alpha_q)^{n-k} \, . \tag{1.13}
$$

This $p$-value can be used to evaluate individual queries or the full redescription $(p, q)$ by setting the query to $p \wedge q$.

It is sometimes convenient to consider only a subset of the entities when evaluating a redescription. Let $\mathcal{E}'$ be a subset of $\mathcal{E}$. The *redescription* $(p, q)$ *conditional to* $\mathcal{E}'$, denoted $(p \sim q \mid \mathcal{E}')$, is evaluated only over the entities in $\mathcal{E}'$. A conditional redescription $(p \sim q \mid \mathcal{E}')$ is *exact* if $\text{supp}(p) \cap \mathcal{E}' = \text{supp}(q) \cap \mathcal{E}'$, that is, the supports of the queries agree in $\mathcal{E}'$. Often, the entities in $\mathcal{E}'$ are selected to be the support of some query. Denoting this query by $r$, we write $(p \sim q \mid r)$ as a shorthand to $(p \sim q \mid \text{supp}(r))$. When using the Jaccard distance, we can identify $(p \sim q \mid r)$ with $(p \wedge e \sim q \wedge r)$; though this identity does not necessarily hold with other distance functions.

## 1.2.5  Distance Functions: Why Jaccard?

We measure the similarity of two descriptions using the Jaccard distance $J(p, q)$ from (1.5). The use of Jaccard can be motivated in many ways. Ramakrishnan et al. (2004) motivated their choice of Jaccard via an argument to *entropy distance*. The supports $\text{supp}(p)$ and $\text{supp}(q)$ can be identified with random variables $X$ and $Y$, respectively, where $X$ chooses elements from $\text{supp}(p)$ uniformly at random (i.e. $\Pr(e \in X : e \in \text{supp}(p)) = 1/|\text{supp}(p)|$ and $\Pr(e \in X : e \notin \text{supp}(p)) = 0$), and $Y$ chooses elements from $\text{supp}(q)$ similarly uniformly at random. The *entropy distance* between $X$ and $Y$ is

$$
D_H(X, Y) = 1 - \frac{I(X; Y)}{H(X, Y)} \, , \tag{1.14}
$$

where $I(X; Y)$ is the *mutual information* of $X$ and $Y$, and $H(X, Y)$ is the *joint entropy* of $X$ and $Y$. In the standard set-theoretic interpretation of information theory (Reza 1961), the mutual information corresponds to the intersection of sets $X$ and $Y$ and the joint entropy to the union of $X$ and $Y$, and hence, (1.14) corresponds to the Jaccard distance (1.5). In particular, it is clear that if $J(p, q) = 0$, then $D_H(X, Y) = 0$ as well.

This motivation is particularly compelling when using algorithms based on decision tree induction (see Sect. 2.2) with the *information gain* splitting criteria (see, e.g. Aggarwal 2015, p. 297).

Another motivation for the Jaccard distance comes from association rule mining (see, e.g. Aggarwal 2015, Chapter 4). If all of the attributes are Boolean and the query language limits the descriptions to monotone conjunctive queries, a redescription $(p, q)$ can be seen as a bidirectional association rule: $p \Rightarrow q$ and $q \Rightarrow p$. A standard measure for the quality of an association rule is its *confidence*: $\mathrm{conf}(p \Rightarrow q) = \mathrm{supp}(p \wedge q) / \mathrm{supp}(p)$, and in case of a redescription, we would like to have a high confidence on both $p \Rightarrow q$ and $q \Rightarrow p$.

Using the notation from (1.8), we can write the confidences as $\mathrm{conf}(p \Rightarrow q) = |\mathcal{E}_{11}| / (|\mathcal{E}_{11}| + |\mathcal{E}_{10}|)$ and $\mathrm{conf}(q \Rightarrow p) = |\mathcal{E}_{11}| / (|\mathcal{E}_{11}| + |\mathcal{E}_{01}|)$. The Jaccard similarity index (or similarity coefficient) can be written as $J(p, q) = |\mathcal{E}_{11}| / (|\mathcal{E}_{11}| + |\mathcal{E}_{10}| + |\mathcal{E}_{01}|)$. It is easy to see that the Jaccard similarity index is never more than either of the confidences. Hence, if $p \sim_\tau q$, then we know that

$$\min\{\mathrm{conf}(p \Rightarrow q), \mathrm{conf}(q \Rightarrow p)\} \geq 1 - \tau .$$

Naturally, our argument carries over to other types of attributes and broader query languages, as we are only operating on the support sets. But in the case of Boolean attributes and monotone conjunctive queries, we can also motivate the use of the Jaccard from the point of view of computational efficiency. It turns out that we can find all association rules with the higher-than-defined Jaccard similarity index efficiently using the min-wise hashing trick (see Aggarwal 2015, Section 4.5.6). These association rules are the redescriptions in the monotone conjunctive query language. Jaccard also has other convenient properties regarding the computation of the redescriptions, as we will see in Sect. 2.3.

We motivated the use of Jaccard distance above by the fact that it ensures good confidence on the association rules. It is natural to ask whether we could consider a mean of the association confidences instead of the minimum. Taking the *harmonic mean* of the confidences, for example, we obtain

$$2 \left( \frac{|\mathcal{E}_{11}| + |\mathcal{E}_{10}|}{|\mathcal{E}_{11}|} + \frac{|\mathcal{E}_{11}| + |\mathcal{E}_{01}|}{|\mathcal{E}_{11}|} \right)^{-1} = \frac{2|\mathcal{E}_{11}|}{2|\mathcal{E}_{11}| + |\mathcal{E}_{10}| + |\mathcal{E}_{01}|} ,$$

which is the famous $F_1$-score (or *Sørensen–Dice coefficient*). It is also very close to the Jaccard similarity coefficient,[3] the only difference being that the shared elements

---

[3] Indeed, we get the Jaccard similarity via simple transformation $J = F_1 / (2 - F_1)$.

are weighted as being twice as important. Unlike the Jaccard, however, the distance based on the $F_1$-score (i.e. $1 - F_1$) is *not* a metric, as it does not satisfy the triangle inequality. While being a metric is not strictly necessary for this purpose, it is often beneficial, and we can motivate the use of Jaccard also as a metricized variant of the harmonic mean of the association confidences.

Another common way of taking the mean of the confidences is the *geometric mean*. This yields

$$\sqrt{\frac{|\mathcal{E}_{11}|}{|\mathcal{E}_{11}| + |\mathcal{E}_{10}|} \frac{|\mathcal{E}_{11}|}{|\mathcal{E}_{11}| + |\mathcal{E}_{01}|}} = \frac{|\mathcal{E}_{11}|}{\sqrt{|\mathcal{E}_{11}| + |\mathcal{E}_{10}|}\sqrt{|\mathcal{E}_{11}| + |\mathcal{E}_{01}|}} \,,$$

that is, the *cosine similarity*. Cosine similarity has an appealing interpretation, as it is the cosine of the angle between the characteristic vectors of $\mathrm{supp}(p)$ and $\mathrm{supp}(q)$. The related cosine distance is not a metric, although the closely related *angular distance* is. Assuming nonnegative vectors, the angular distance is defined as

$$\cos^{-1}\left(\frac{|\mathcal{E}_{11}|}{\sqrt{|\mathcal{E}_{11}| + |\mathcal{E}_{10}|}\sqrt{|\mathcal{E}_{11}| + |\mathcal{E}_{01}|}}\right)/\pi \,.$$

The third common mean is the standard *arithmetic mean*, yielding

$$\frac{1}{2}\left(\frac{|\mathcal{E}_{11}|}{|\mathcal{E}_{11}| + |\mathcal{E}_{10}|} + \frac{|\mathcal{E}_{11}|}{|\mathcal{E}_{11}| + |\mathcal{E}_{01}|}\right) = \frac{|\mathcal{E}_{11}|(2|\mathcal{E}_{11}| + |\mathcal{E}_{10}| + |\mathcal{E}_{01}|)}{2(|\mathcal{E}_{11}| + |\mathcal{E}_{10}|)(|\mathcal{E}_{11}| + |\mathcal{E}_{01}|)} \,.$$

This mean is arguably the least interesting of the three. If we denote the Jaccard similarity by $J$, the $F_1$-score by $F_1$, the cosine similarity by $C$, and the arithmetic mean by $A$, then, by the inequality of arithmetic and geometric means, we have that

$$J \leq F_1 \leq C \leq A \,, \tag{1.15}$$

with the first inequality being strict unless $|\mathcal{E}_{11}| = 0$ or $|\mathcal{E}_{10}| = |\mathcal{E}_{01}| = 0$. The last two inequalities are strict unless $|\mathcal{E}_{11}| = 0$ or $|\mathcal{E}_{10}| = |\mathcal{E}_{01}|$.

The means of the association confidences are not the only possible alternatives to the Jaccard. There has been an extensive study of different interestingness measures for association rules (see, e.g. Geng and Hamilton 2006), and in principle, any symmetric interestingness measure could be used as the basis for the distance between the redescriptions. But given the many benefits of the Jaccard distance—and the fact that all current redescription mining algorithms aim at optimizing it—there would have to be very strong reasons to use other distances.

## *1.2.6  Sets of Redescriptions*

Redescription mining, as defined in Definition 9, is an exhaustive enumeration task, the goal being to output *all* valid redescriptions that satisfy the constraints. This is a common approach in data mining (cf. frequent itemset and subgraph mining), but it can yield many redundant redescriptions. For example, consider again the ecologist using redescription mining to find bioclimatic niches for mammals. The top redescriptions she found from the data covering Europe are listed in Table 1.1.

All redescriptions in Table 1.1 have just one mammal, the polar bear, in query $p$. Query $q$ always describes cold environments using different months' temperatures. In summary, these five redescriptions all describe the same phenomenon: polar bears can be found in cold environments (in the northern hemisphere). It is clear that we do not need all of these redescriptions, and some pruning would be beneficial.

Some pruning has, in fact, been done already. These five redescriptions are not the only possible ones; any conjunction or disjunction of the right-hand side queries $q$ would yield equally good results in this data set. But as the results would not be any better, redescription mining algorithms usually apply *Occam's razor* and report only the simplest redescriptions that achieve the same quality.

The pruning of overly complex redescriptions still does not solve the problem of redundant redescriptions, though. Indeed, all the quality measures and constraints discussed above consider each redescription separately, and they do not consider the final *set* of redescriptions.

The simplest way to reduce redundancy is to consider either the supports of the redescriptions or their attributes. When considering the attributes, we can limit the number of times an attribute can appear in different queries (Ramakrishnan et al. 2004). The problem with this approach is that we need to remove the often-used attributes from the set of attributes during the mining, thus creating the risk that we will not find some good redescription because we have removed an attribute that was vital to it earlier in the mining process.

Galbrun and Miettinen (2012c) propose a simple support-based filtering to remove redundant redescriptions. In this scheme, we first order all (valid) redescriptions descending in their similarity. We then take the topmost redescription, move

**Table 1.1** Redundant redescriptions in the bioclimatic niche finding setting

| $p$ | $\sim$ | $q$ |
| --- | --- | --- |
| polar bear | $\sim$ | $[-7.1 \le t_5 \le -3.4]$ |
| polar bear | $\sim$ | $[-16.7 \le t_3 \le -11.5]$ |
| polar bear | $\sim$ | $[-4.5 \le t_{10}^+ \le -1.0]$ |
| polar bear | $\sim$ | $[1.0 \le t_9^+ \le 3.5]$ |
| polar bear | $\sim$ | $[-9.6 \le t_4^+ \le -5.6]$ |

Variables $t_n$ stand for average temperature in month $n$, while variables $t_n^+$ stand for the maximum temperature in month $n$, both in degrees Celsius. The example is adapted from Kalofolias et al. (2016)

it to the list of non-redundant redescriptions, and mark the entities in its support 'used'. We can then re-evaluate the remaining redescriptions, but only taking into account the non-used entities. All redescriptions that are deemed invalid (e.g. their support size becomes too low or their distance too high) are considered redundant and removed. The process is then re-run with the remaining redescriptions and entities. The process ends when either the list of redescriptions or the set of entities becomes empty, at which point only the redescriptions in the list of non-redundant redescriptions are returned to the user.

This approach can prune also interesting redescriptions as all redescriptions with the same entities in the support are considered redundant to each other, even if their attributes are completely different. A more fine-grained alternative, proposed for itemset mining by Gallo et al. (2007), is to consider the *rectangles* $\text{rect}(p, q) = \text{supp}(p, q) \times \text{att}(p, q)$, that is, $\text{rect}(p, q)$ is a set of entity–attribute pairs $\text{rect}(p, q) = \{(e, a) : e \in \text{supp}(p, q), a \in \text{att}(p, q)\}$. Ordering the redescriptions again based on their similarity, we consider redundant those redescriptions $(p, q)$ for which $|\text{rect}(p, q) \cap \text{rect}(p', q')| / |\text{rect}(p', q')| > \theta$ for some higher-similarity redescription $(p', q')$ and for some predetermined threshold $\theta$.

Another alternative for filtering out redundant redescriptions was proposed by Kalofolias et al. (2016). Their approach is based on the formalization of *surprisingness* (or *interestingness*), using the likelihood of the result under a constrained maximum-entropy distribution (De Bie 2011). The idea is to build a probability distribution for the data, allowing us to determine the *likelihood* of seeing a particular redescription with a particular support size if the data were a random sample from that distribution. If this likelihood is high, we consider the redescription unsurprising, and hence redundant; vice versa, redescriptions with low likelihood are considered surprising. The crux of this approach is the distribution: it should have maximum entropy over all distributions under which those redescriptions that we have already considered surprising are certain. To that end, the distribution is updated every time we find a new surprising redescription. The update is done by adding a constraint that limits the values the distribution can take in the just-seen redescription's rectangle (to make sure the redescription becomes certain under the new distribution). The distribution of the data is then added to the maximum-entropy distribution that admits this (and previous) constraints.

The approach of Kalofolias et al. (2016) facilitates a principled way of removing redundant (or unsurprising) redescriptions. We can first add the highest-quality redescription (determined, e.g. by the similarity) as a constraint and sort the remaining redescriptions ascending on their likelihood. After adding the least-likely (i.e. most surprising) redescription, we update the distribution and the likelihoods and re-order the remaining redescriptions. We do not have to prune out the redescriptions, as the uninteresting ones are just pushed to the bottom of the list.

The problem with the maximum-entropy based approach is that updating the distribution is computationally very expensive (especially if the redescriptions involve complex queries with many variables). This means that the ordering cannot be done in real-time, and the likelihood cannot be used as a search or pruning criterion for the redescription mining algorithms. Nonetheless, it does provide an appealing way of turning a set of good redescriptions into a good set of redescriptions.

A different approach for removing the redundant redescriptions was proposed by van Leeuwen and Galbrun (2015). Their approach is inspired by the *minimum description length* (MDL) principle of Rissanen (1978). van Leeuwen and Galbrun (2015) consider the *translation rules* between two data sets. The translation rules are essentially association rules between the data sets, and van Leeuwen and Galbrun (2015) consider translations to both directions. When a translation rule applies in both directions, it can be considered a monotone conjunctive redescription over Boolean attributes. Given two data tables $D_1$ and $D_2$, the goal of van Leeuwen and Galbrun is to find the set of translation rules (i.e. association rules and redescriptions) that minimizes the total number of bits needed to encode (1) the translation rules themselves, (2) the corrections that are needed in order to build $D_2$ given $D_1$ and the translation rules, and (3) the corrections that are needed to build $D_1$ given $D_2$ and the translation rules. With the translation rules and corrections, one can rebuild one data set if the other is known.

The MDL-inspired approach of van Leeuwen and Galbrun (2015) seeks a balance between having as few translation rules as possible and having as few corrections as possible. The encoding also takes into account how well the entities are covered by the supports of the translation rules. Those entity–attribute pairs that are present in the data but are not covered by any of the translation rules must be encoded in the corrections; everything else being equal, this increases the encoding length.

## 1.3   Related Data Mining Problems

Association rule mining (Agrawal et al. 1993) is one of the classical problems in data mining. It can be seen as a precursor of redescription mining, with the latter allowing for more complex descriptions and focusing on equivalences instead of implications (Ramakrishnan et al. 2004). That said, many redescription mining algorithms draw ideas and inspiration from association rule mining. This is especially true when the query language is restricted to monotone conjunctive queries over Boolean attributes, in which case one can use algorithms for association rule mining, closed itemset mining, or formal concept analysis (Ganter and Wille 1999) almost directly (see Sect. 2.1).

Another classical problem that has strongly influenced redescription mining is classification. Let query $q$ be fixed; our goal is to find query $p$. This can be seen as a binary classification problem. Our data are the entities and their attributes that do not belong to views($q$). The class labels are given by supp($q$): the label for entities $e \in$ supp($q$) is 1, and the label for entities $e \notin$ supp($q$) is 0. Finding a classifier with good precision and recall is now essentially equivalent to finding a query $p$ that is close to $q$ in the Jaccard distance. For the classifier to be a proper description, it must come from our query language $\mathcal{Q}$; in practice, the query language can be defined so that it matches the classifiers. More limiting to the use of various classification algorithms in redescription mining is the common aim of having interpretable descriptions. This makes otherwise successful classification methods such as kernel support vector machines (Cortes and Vapnik 1995) or deep belief networks (Hinton et al. 2006) less appealing for redescription mining. Decision tree induction, on the other hand, is a common approach for mining redescriptions (see Sect. 2.2).

In subgroup discovery (Wrobel 1997), the input contains features and a target variable over observations, and the goal is to find queries that describe groups that have 'interesting' behaviour in the target variable. What is considered interesting is, of course, application-dependant, but the found subgroups are often assumed to have different statistical properties (e.g. average) in the target variable when compared to the rest of the observations. Exceptional model mining (Leman et al. 2008) extends subgroup discovery by replacing the target variable with a target model; now, the interesting subgroups are those that violate the target model. The supervised nature of subgroup discovery and exceptional model mining as well as their concentration on the *exceptional* subgroups distinguishes them from redescription mining. It could be argued, though, that redescription mining also aims for exceptional subgroups: if the support of $p$ is not in any way exceptional in the other views, there will be no way to build the query $q$ with high similarity. In that sense, redescription mining can be seen as an unsupervised version of subgroup discovery or exceptional model mining.

In constraint programming (Rossi et al. 2006), a task is formulated by specifying the constraints a solution must satisfy in order to be acceptable. De Raedt et al. (2010) first modelled the itemset mining task as a constraint programming problem, and Guns et al. (2013) later proposed a formulation in this framework for several pattern mining tasks, including the task of mining exact conjunctive redescriptions.

Clustering is a classical unsupervised data analysis method with the goal of grouping the entities in such a way that entities in the same group are as similar to each other as possible, and the objects in different groups are as dissimilar from each other as possible. In subspace clustering, the similarity of the objects is only calculated over a subset of the attributes (for more information, see, e.g. Kröger and Zimek 2009, and references therein). Biclustering (Madeira and Oliveira 2004) is a related method where we simultaneously cluster the objects based on a subset of attributes, and the attributes based on a subset of objects. A query $q$ can be interpreted to select a subset of the attributes, namely att($q$), and a group of objects, supp($q$), that are in some sense 'similar' to each other. This similarity, however, should not be understood in the classical sense (e.g. as an Euclidean distance), as the query can allow a wide range of values for different attributes. Furthermore, if the query contains disjunctions, two objects in supp($q$) do not have to agree in any attribute. But while the general queries do not make good biclusters, the connection can be utilized the other way around, using subspace or biclustering algorithms to find good queries. CLIQUE (Agrawal et al. 1998) is an example of a subspace clustering algorithm that finds redescription-style queries as the minimal descriptions of the clusters, while Jin et al. (2008) use biclusters to build redescriptions.

The above methods mostly concentrate on finding one query, but the distinctive characteristics of redescription mining are the 'two views' it provides from the pair of queries. Indeed, redescription mining is an example of *multi-view data mining methods*. Other examples include, but are not limited to, *multi-view clustering* (Bickel and Scheffer 2004), where the attributes are divided into two views and the clustering is done separately over each view; *multi-view subgroup discovery* (Umek et al. 2009), where the subgroup discovery is done over multiple views; and various *multi-view matrix and tensor factorization* methods (e.g. Miettinen 2012; Gupta et al. 2013; Khan and Kaski 2014), which use (partially) the same factors to decompose multiple matrices or tensors.

Let us look at the last group of methods more carefully. In its simplest form, the goal of multi-view matrix factorization is to factorize two matrices, $A$ and $B$, both having the same number of rows, as $A \approx XY$ and $B \approx XZ$. The relation to redescription mining is easy to see when we restrict all matrices to be binary and consider the rank-1 factorizations $A \approx xy^T$ and $B \approx xz^T$. Now, the entries of vectors $y$ and $z$ that are 1 select some columns of $A$ and $B$, respectively, and the non-zero entries of $x$ select the rows of the matrices. In the rows selected by $x$, matrices $A$ and $B$ should have 1 in all columns selected by $y$ or $z$, respectively. Hence, the vectors $x$ and $y$ correspond to conjunctive queries over the columns of $A$ and $B$, while vector $x$ corresponds to the support of these queries. Unlike redescription mining, however, in this setting, vector $x$ is chosen by the algorithm and does not have to correspond to all rows where the queries hold; indeed, $x$ can also select rows where the queries do not hold. The goal of the factorization is not to find queries with similar support, but to find the query and the support that minimize the *reconstruction error*.

## 1.4  A Short History

Redescription mining was first formalized by Ramakrishnan et al. (2004). Their algorithm, CARTwheels, was based on the idea of alternatively growing decision trees over one data table with only Boolean attributes. After the seminal work of Ramakrishnan et al., the work on redescription mining continued to concentrate on Boolean data. Zaki and Ramakrishnan (2005) studied exact and conditional redescriptions over Boolean data. They concentrated only on conjunctive queries and presented a way to use existing closed itemset mining algorithms for exhaustively enumerating all exact conjunctive redescriptions over Boolean data. Parida and Ramakrishnan (2005) studied the theory of exact redescriptions over Boolean attributes, presenting general frameworks for mining all redescriptions where the queries are pure conjunctions, whether in monotone conjunctive normal form or monotone disjunctive normal form.

Linearly parsable queries over Boolean attributes were introduced by Gallo et al. (2008). Their general approach was extended to numerical and categorical attributes by Galbrun and Miettinen (2012b) in an algorithm called ReReMi. Decision-tree-based methods for arbitrary data types were introduced by Zinchenko et al. (2015), who also studied how well the redescriptions predict the unseen data. In a similar manner, Mihelčić et al. (2016) used predictive clustering trees for mining redescriptions and Mihelčić et al. (2017) extended that approach to random forests.

The Siren tool was developed for mining, visualizing, and interacting with redescriptions (Galbrun and Miettinen 2012a,c, 2014). Later, Mihelčić and Šmuc (2016) proposed a tool called InterSet for visualizing and working with *sets* of redescriptions.

Finding redundant redescriptions has been a problem since the begin of redescription mining. Ramakrishnan et al. (2004) allow each attribute to appear in only a predefined number of redescriptions before being removed from the set of attributes. Zaki and Ramakrishnan (2005) studied the minimal generators of exact redescriptions over Boolean variables and showed how to mine non-redundant exact redescriptions from the minimal generators.

Galbrun and Miettinen (2012c) proposed a way to prune the redundant redescriptions based on the support of already-found redescriptions. This idea was further extended by Mihelčić et al. (2017). Other methods for removing redundant redescriptions were proposed by van Leeuwen and Galbrun (2015) (for monotone conjunctive queries over Boolean attributes using MDL) and Kalofolias et al. (2016) (for general queries over arbitrary attributes, using maximum-entropy distributions).

Redescription mining has been applied in various domains, including bioinformatics (e.g. Kumar 2007; Ramakrishnan and Zaki 2009; Gaidar 2015), electrical engineering (Goel et al. 2010), and political sciences (Galbrun and Miettinen 2016), to name a few. We will discuss some of these applications in Sect. 3.1.

In addition to standard redescription mining, various extensions and variants of redescription mining have been proposed over the years. *Relational redescription mining* (Galbrun and Kimmig 2014) extends redescription mining to relational data, finding ways to describe groups of entities based on their individual properties and the relations between them. Ramakrishnan et al. (2004) proposed *storytelling* as a method for connecting different entities via chains of redescriptions. We will discuss these variants in more details respectively in Sects. 3.2 and 3.3.

# References

Aggarwal CC (2015) Data Mining: The Textbook. Springer, Cham, https://doi.org/10.1007/978-3-319-14142-8

Agrawal R, Imielinski T, Swami A (1993) Mining association rules between sets of items in large databases. In: Proceedings of the 1993 ACM SIGMOD International Conference on Management of Data (SIGMOD'93), pp 207–216, https://doi.org/10.1145/170035.170072

Agrawal R, Gehrke J, Gunopulos D, Raghavan P (1998) Automatic subspace clustering of high dimensional data for data mining applications. SIGMOD Rec 27(2):94–105, https://doi.org/10.1145/276305.276314

Bickel S, Scheffer T (2004) Multi-view clustering. In: Proceedings of the 4th IEEE International Conference on Data Mining (ICDM'04), pp 19–26, https://doi.org/10.1109/ICDM.2004.10095

Cortes C, Vapnik V (1995) Support-vector networks. Mach Learn 20(3):273–297, https://doi.org/10.1007/BF00994018

De Bie T (2011) Maximum entropy models and subjective interestingness: an application to tiles in binary databases. Data Min Knowl Discov 23(3):407–446, https://doi.org/10.1007/s10618-010-0209-3

De Raedt L, Guns T, Nijssen S (2010) Constraint programming for data mining and machine learning. In: Proceedings of the 24th AAAI Conference on Artificial Intelligence (AAAI'10)

Gaidar D (2015) Mining redescriptors in Staphylococcus aureus data. Master's thesis, Universität des Saarlandes, Saarbrücken

Galbrun E (2013) Methods for redescription mining. PhD thesis, Department of Computer Science, University of Helsinki

Galbrun E, Kimmig A (2014) Finding relational redescriptions. Mach Learn 96(3):225–248, https://doi.org/10.1007/s10994-013-5402-3

Galbrun E, Miettinen P (2012a) A case of visual and interactive data analysis: Geospatial redescription mining. In: Proceedings of the ECML PKDD 2012 Workshop on Instant and Interactive Data Mining (IID'12), URL http://adrem.ua.ac.be/iid2012/papers/galbrun_miettinen-visual_and_interactive_geospatial_redescription_mining.pdf, accessed 25 Oct 2017.

Galbrun E, Miettinen P (2012b) From black and white to full color: Extending redescription mining outside the Boolean world. Stat Anal Data Min 5(4):284–303, https://doi.org/10.1002/sam.11145

Galbrun E, Miettinen P (2012c) Siren: An interactive tool for mining and visualizing geospatial redescriptions [demo]. In: Proceedings of the 18th ACM SIGKDD International Conference on Knowledge Discovery and Data Mining (KDD'12), pp 1544–1547, https://doi.org/10.1145/2339530.2339776

Galbrun E, Miettinen P (2014) Interactive redescription mining. In: Proceedings of the 2014 ACM SIGMOD International Conference on Management of Data (SIGMOD'14), pp 1079–1082, https://doi.org/10.1145/2588555.2594520

Galbrun E, Miettinen P (2016) Analysing political opinions using redescription mining. In: IEEE International Conference on Data Mining Workshops, pp 422–427, https://doi.org/10.1109/ICDMW.2016.0066

Gallo A, De Bie T, Cristianini N (2007) MINI: Mining informative non-redundant itemsets. In: Proceedings of the 11th European Conference on Principles and Practice of Knowledge Discovery in Databases (PKDD'07), pp 438–445

Gallo A, Miettinen P, Mannila H (2008) Finding subgroups having several descriptions: Algorithms for redescription mining. In: Proceedings of the 8th SIAM International Conference on Data Mining (SDM'08), pp 334–345, https://doi.org/10.1137/1.9781611972788.30

Ganter B, Wille R (1999) Formal Concept Analysis: Mathematical Foundations. Springer, Berlin, https://doi.org/10.1007/978-3-642-59830-2

Geng L, Hamilton HJ (2006) Interestingness measures for data mining: A survey. ACM Comput Surv 38(3):Article 9, https://doi.org/10.1145/1132960.1132963

Goel N, Hsiao MS, Ramakrishnan N, Zaki MJ (2010) Mining complex Boolean expressions for sequential equivalence checking. In: Proceedings of the 19th IEEE Asian Test Symposium (ATS'10), pp 442–447, https://doi.org/10.1109/ATS.2010.81

Grinnell J (1917) The niche-relationships of the california thrasher. The Auk 34(4):427–433

Guns T, Nijssen S, De Raedt L (2013) k-Pattern set mining under constraints. IEEE Trans Knowl Data En 25(2):402–418, https://doi.org/10.1109/TKDE.2011.204

Gupta SK, Phung D, Adams B, Venkatesh S (2013) Regularized nonnegative shared subspace learning. Data Min Knowl Disc 26(1):57–97, https://doi.org/10.1007/s10618-011-0244-8

Hinton GE, Osindero S, Teh YW (2006) A fast learning algorithm for deep belief nets. Neural Comput 18(7):1527–1554, https://doi.org/10.1162/neco.2006.18.7.1527

Inselberg A (2009) Parallel Coordinates: Visual Multidimensional Geometry and Its Applications. Springer, Dordrecht, https://doi.org/10.1007/978-0-387-68628-8

Jin Y, Murali TM, Ramakrishnan N (2008) Compositional mining of multirelational biological datasets. ACM Trans Knowl Disc Data 2(1):2–35, https://doi.org/10.1145/1342320.1342322

Kalofolias J, Galbrun E, Miettinen P (2016) From sets of good redescriptions to good sets of redescriptions. In: Proceedings of the 16th IEEE International Conference on Data Mining (ICDM'16), pp 211–220, https://doi.org/10.1109/ICDM.2016.0032

Khan SA, Kaski S (2014) Bayesian multi-view tensor factorization. In: Proceedings of the 2014 European Conference on Machine Learning and Principles and Practice of Knowledge Discovery in Databases (ECML-PKDD'14), pp 656–671, https://doi.org/10.1007/978-3-662-44848-9_42

Kröger P, Zimek A (2009) Subspace clustering techniques. In: Liu L, Özsu MT (eds) Encyclopedia of Database Systems, Springer, New York, pp 2873–2875, https://doi.org/10.1007/978-0-387-39940-9_607

Kumar D (2007) Redescription mining: Algorithms and applications in bioinformatics. PhD thesis, Department of Computer Science, Virginia Polytechnic Institute and State University

van Leeuwen M, Galbrun E (2015) Association discovery in two-view data. IEEE Trans Knowl Data Eng 27(12):3190–3202, https://doi.org/10.1109/TKDE.2015.2453159

Leman D, Feelders A, Knobbe AJ (2008) Exceptional model mining. In: Proceedings of the 2008 European Conference on Machine Learning and Principles and Practice of Knowledge Discovery in Databases (ECML-PKDD'08), vol 5212, pp 1–16, https://doi.org/10.1007/978-3-540-87481-2_1

Madeira SC, Oliveira AL (2004) Biclustering algorithms for biological data analysis: A survey. IEEE Trans Comput Bio Bioinform 1(1):24–45, https://doi.org/10.1109/TCBB.2004.2

Miettinen P (2012) On finding joint subspace boolean matrix factorizations. In: SIAM International Conference on Data Mining (SDM'12), pp 954–965, https://doi.org/10.1137/1.9781611972825.82

Mihelčić M, Šmuc T (2016) InterSet: Interactive redescription set exploration. In: Proceedings of the 19th International Conference on Discovery Science (DS'16), vol 9956, pp 35–50

Mihelčić M, Džeroski S, Lavrač N, Šmuc T (2017) A framework for redescription set construction. Expert Syst Appl 68:196–215, https://doi.org/10.1016/j.eswa.2016.10.012

Mihelčić M, Džeroski S, Lavrač N, Šmuc T (2016) Redescription mining with multi-target predictive clustering trees. In: Proceedings of the 4th International Workshop on the New Frontiers in Mining Complex Patterns (NFMCP'15), pp 125–143, https://doi.org/10.1007/978-3-319-39315-5_9

Mihelčić M, Džeroski S, Lavrač N, Šmuc T (2017) Redescription mining augmented with random forest of multi-target predictive clustering trees. J of Intell Inf Syst pp 1–34, https://doi.org/10.1007/s10844-017-0448-5

Parida L, Ramakrishnan N (2005) Redescription mining: Structure theory and algorithms. In: Proceedings of the 20th National Conference on Artificial Intelligence and the 7th Innovative Applications of Artificial Intelligence Conference (AAAI'05), pp 837–844

Ramakrishnan N, Zaki MJ (2009) Redescription mining and applications in bioinformatics. In: Chen J, Lonardi S (eds) Biological Data Mining, Chapman and Hall/CRC, Boca Raton, FL

Ramakrishnan N, Kumar D, Mishra B, Potts M, Helm RF (2004) Turning CARTwheels: An alternating algorithm for mining redescriptions. In: Proceedings of the 10th ACM SIGKDD International Conference on Knowledge Discovery and Data Mining (KDD'04), pp 266–275, https://doi.org/10.1145/1014052.1014083

Reza FM (1961) An Introduction to Information Theory. McGraw-Hill, New York

Rissanen J (1978) Modeling by shortest data description. Automatica 14(5):465–471, https://doi.org/10.1016/0005-1098(78)90005-5

Rossi F, Van Beek P, Walsh T (2006) Handbook of constraint programming. Elsevier, Amsterdam

Soberón J, Nakamura M (2009) Niches and distributional areas: Concepts, methods, and assumptions. Proc Natl Acad Sci USA 106(Supplement 2):19,644–19,650, https://doi.org/10.1073/pnas.0901637106

Umek L, Zupan B, Toplak M, Morin A, Chauchat JH, Makovec G, Smrke D (2009) Subgroup discovery in data sets with multi-dimensional responses: A method and a case study in traumatology. In: Proceedings of the 12th Conference on Artificial Intelligence in Medicine (AIME'09), vol 5651, pp 265–274, https://doi.org/10.1007/978-3-642-02976-9_39

Wrobel S (1997) An algorithm for multi-relational discovery of subgroups. In: Proceedings of the First European Symposium on Principles of Data Mining and Knowledge Discovery (PKDD'97), vol 1263, pp 78–87, https://doi.org/10.1007/3-540-63223-9_108

Zaki MJ, Ramakrishnan N (2005) Reasoning about sets using redescription mining. In: Proceedings of the 11th ACM SIGKDD International Conference on Knowledge Discovery and Data Mining (KDD'05), pp 364–373, https://doi.org/10.1145/1081870.1081912

Zinchenko T, Galbrun E, Miettinen P (2015) Mining predictive redescriptions with trees. In: IEEE International Conference on Data Mining Workshops, pp 1672–1675, https://doi.org/10.1109/ICDMW.2015.123

# Chapter 2
# Algorithms for Redescription Mining

As explained in the previous chapter, the aim of redescription mining is to find valid redescriptions for given data, query language, similarity relation, and user-specified constraints. In other words, we need to explore the search space consisting of query pairs from the query language, looking for those pairs that have similar enough support in the data and that satisfy the other constraints.

The input to the algorithms discussed in this chapter takes the form of data tables, as described in Definition 3. Therefore, our discussion of algorithmic techniques for mining redescriptions focuses on this simpler model. In fact, the presentation of most algorithms will assume that the data consist of two tables, $\mathbf{D}_1$ and $\mathbf{D}_2$, since it is the setting most commonly encountered and because it will make the different algorithms more readily comparable. Still, as we will see, most of the algorithms can be adapted to accommodate an arbitrary number of views. Throughout this chapter, the similarity of descriptions is evaluated using the Jaccard index as defined in (1.4). Indeed, the Jaccard index is used for this purpose almost universally by existing redescription mining algorithms.

Depending on the query language, the search space might be very large. In particular, there are $2^{2^k}$ non-equivalent unrestricted Boolean expressions over a set of $k$ predicates. Hence, given a set of $n$ predicates, there are

$$\kappa_n = \sum_{k=0}^{n} \binom{n}{k} 2^{2^k}$$

different expressions of arbitrary length. Therefore, when looking at two Boolean data tables with $n_1 = 2|\text{att}(\mathbf{D}_1)|$ and $n_2 = 2|\text{att}(\mathbf{D}_2)|$ predicates, respectively (all attributes as well as their negations), there are potentially up to $(\kappa_{n_1} - 1)(\kappa_{n_2} - 1)$ pairs of non-empty queries to examine. In the presence of non-Boolean attributes, the number of predicates that can be constructed might be extremely large, and the number of query pairs will be even larger. For reasons of interpretability, one would generally only consider queries involving at most a small fixed number of

© The Author(s) 2017      25
E. Galbrun, P. Miettinen, *Redescription Mining*, SpringerBriefs
in Computer Science, https://doi.org/10.1007/978-3-319-72889-6_2

**Fig. 2.1** Classification of
redescription mining
algorithms

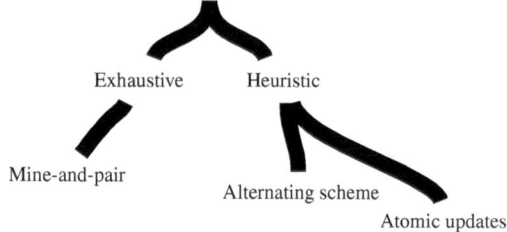

predicates and would impose syntactic restrictions on the combination of predicates, significantly reducing the amount of candidate pairs. Still, the search space of query pairs generally remains very large, and we need efficient strategies for exploring it. In this chapter, we present the different strategies that have been proposed for mining redescriptions, that is, we look in more detail at the various algorithmic approaches available for the task.

As mentioned in Sect. 1.3, redescriptions are strongly related to two other data mining tasks: association rule mining and classification. These two tasks provide basic techniques that have been adapted to develop algorithms for mining redescriptions. On the one hand, association rule mining inspired algorithms that first mine queries separately from the different views before combining the obtained queries across the views into redescriptions. On the other hand, the fact that building one query of a redescription when the other query is fixed corresponds to a classification task has yielded another family of algorithms: iterative algorithms that alternate between the views. These algorithms derive target labels from a query obtained at a previous iteration and use classification techniques, typically decision tree induction, to build a matching query in the next iteration. A third approach for mining redescriptions consists in growing them greedily. In this approach, the queries are extended progressively through atomic updates, such as appending new literals to either query, always trying to improve the quality of the redescription.

The proposed algorithms can be divided between exhaustive and heuristic strategies. Mine-and-pair algorithms based on association rule mining techniques are typically exhaustive. Alternating algorithms based on decision tree induction and algorithms that use atomic updates to grow the queries greedily typically rely on heuristics. This classification of redescription mining algorithms is illustrated in Fig. 2.1.

## 2.1 Finding Queries Using Itemset Mining

The simplest exploration strategy consists of two steps. First, individual queries are mined from the data set independently. Second, queries with similar supports are paired to form redescriptions.

The main advantage of such a mine-and-pair strategy is that it allows to adapt frequent itemset mining algorithms in a very straightforward fashion. On the other hand, because they build on techniques from itemset mining and rely on an exhaustive enumeration, the query language handled with this strategy is typically limited to monotone conjunctive queries over Boolean attributes.

Monotone conjunctive queries, such as $a \wedge b \wedge c$, can be considered as a way to select a subset of predicates (simple Boolean attributes as well as binarized numerical or categorical attributes). If we identify the predicates with items and entities with transactions, monotone conjunctive queries become itemsets, in this case $\{a, b, c\}$ (see e.g. Aggarwal 2015, Chapter 4). In what follows, we denote the monotone conjunctive queries and the corresponding itemsets with capital letters, as in $A = a \wedge b \wedge c = \{a, b, c\}$. The support of an itemset is the same as the support of the corresponding query. A particular feature of Boolean data tables is that we can always switch the roles of attributes and entities, simply by transposing the table. Hence, if $E \subseteq \mathcal{E}$ is a set of entities, we can identify it with a monotone conjunctive query in the transpose of the data. The support of this query, that is, the set of all predicates that are true for all entities in $E$, is denoted by $dscr(E) = \{p : p(e) = true$ for all $e \in \mathcal{E}\}$.

In the literature of frequent itemset mining (FIM) (see e.g. Aggarwal 2015, Section 5.2), an itemset $I$ is said to be *closed* if none of its supersets has the same support, that is, $I$ is closed if and only if $dscr(supp(I)) = I$. In Formal Concept Analysis (FCA) (Ganter and Wille 1999), the pair $(I, E)$ comprising a closed itemset $I$ and its support $E$ is called a *formal concept*. The *itemset lattice* is a fundamental concept in itemset mining. It is the partial ordering of itemsets based on set inclusion relationships, that is, the ordering such that itemset $A$ is less than itemset $B$ whenever $A \subset B$. When restricted to only closed itemsets (formal concepts), such a lattice is known as a *closed itemset lattice* (in FIM) or a *concept lattice* (in FCA). For a closed itemset $I$, a subset $J \subseteq I$ such that $supp(J) = supp(I)$ is called a *generator* of $I$. The generator is said to be *proper* when the inclusion is strict and *minimal* when there is no other generator $J'$ of $I$ such that $J' \subset J$. We denote the set of minimal generators of an itemset $I$ as $\mathcal{M}(I)$.

Over the last couple of decades, a great number of algorithms have been developed to mine monotone conjunctive queries over a fixed set of predicates (see Aggarwal 2015, Section 4.4 for an overview). Typically, they exploit the anti-monotonicity of the support of queries to safely prune the search space, resulting in highly efficient complete enumeration procedures.

In particular, monotone conjunctive redescriptions can be mined exhaustively in a level-wise fashion similar to the *Apriori* algorithm (Agrawal and Srikant 1994; Mannila et al. 1994). The support cardinality of both queries and of their intersection, as well as some associated measures, are anti-monotonic and can be used safely for pruning. However, distance measures such as the Jaccard distance are typically not monotonic, even in this simplest case, and thus do not provide an effective pruning criterion.

If the number of views is small, the most practical approach is to mine queries from each view separately, then to pair them across the views. If the number of views is large, in particular when each attribute is associated to a distinct view, one might instead mine queries over all predicates pooled together, then pair queries with similar supports, provided that they involve attributes from disjoint sets of views.

An alternative to this approach is to replace the pairing step with a splitting step, that is, to pool together all predicates for the mining step, then split the queries depending on views. However, the existence of a query does not imply that it can be split into two subqueries that both hold with the same supports. More generally, there is no guarantee that there will be a way to split the query found into two subqueries over disjoint views with sufficiently similar supports.

### 2.1.1  The MID Algorithm

One algorithm that follows the mine-and-pair strategy is the MID algorithm of Gallo et al. (2008). A sketch of the algorithm is provided in Algorithm 2.1.

With two tables $\mathbf{D}_1$ and $\mathbf{D}_2$ as input, the MID algorithm starts by mining frequent closed itemsets from either table separately (line 2 in Algorithm 2.1). Still considering the two sides separately, the obtained itemsets are combined into more complex queries using conjunctions and disjunctions (line 7), up to a level $\kappa$ chosen by the user. In each iteration, only the $N$ most significant candidate queries, that is, the $N$ candidates with the lowest $p$-values are retained (line 9) to be combined further in the next iteration. The $p$-values are computed as explained in Sect. 1.2.4, using (1.12) and (1.13). Finally, the queries found for either side are combined into pairs, storing those that are sufficiently similar (line 12).

---

**Algorithm 2.1** Sketch of the MID algorithm

---

**Input:** Two Boolean data tables $\mathbf{D}_1$ and $\mathbf{D}_2$, similarity $\sim$, maximum $p$-value $p_{max}$, number of queries to select $N$, and maximum level $\kappa$.
**Output:** Redescriptions $\mathcal{R}$.
 1: **for** side $i \in \{1, 2\}$ **do**
 2:      $\mathcal{Q}_i^{(0)} \leftarrow \{q : q$ is a closed frequent itemset from $\mathbf{D}_i$ or its negation and $p$-value$(q) \leq p_{max}\}$
 3:      $\mathcal{Q}_i^{(1)} \leftarrow N$ queries with the lowest $p$-value from $\mathcal{Q}_i^{(0)}$
 4:      **for** level $k \in \{1, \ldots, (\kappa - 1)\}$ **do**
 5:           $\mathcal{Q}_i^{(k+1)} \leftarrow \mathcal{Q}_i^{(k)}$
 6:           **for** operator $\circ \in \{\wedge, \vee\}$ **do**
 7:                $\mathcal{Q}_i^{(k+1)} \leftarrow \mathcal{Q}_i^{(k+1)} \cup \{q \circ q' : q, q' \in \mathcal{Q}_i^{(k)}, p$-value$(q \circ q') \leq p_{max}\}$
 8:           **end for**
 9:           $\mathcal{Q}_i^{(k+1)} \leftarrow N$ queries with the lowest $p$-value from $\mathcal{Q}_i^{(k+1)}$
10:      **end for**
11: **end for**
12: $\mathcal{R} \leftarrow \{(p, q) \in \mathcal{Q}_1^{(\kappa)} \times \mathcal{Q}_2^{(\kappa)} : p \sim q\}$
13: **return** $\mathcal{R}$

---

## 2.1.2  Mining Redescriptions with the CHARM-L Algorithm

Zaki and Hsiao (2005) introduced a frequent itemset mining algorithm that operates on the closed itemset lattice, called CHARM, as well as its variant that explicitly constructs the lattice, called CHARM-L. Zaki and Ramakrishnan (2005) then developed a method to extract all minimal exact conditional redescriptions from the lattice returned by CHARM-L.

Recall that a *conditional redescription* is an expression $(p \sim q \mid r)$ where $p$, $q$ and $r$ are queries over disjoint sets of attributes, with at least $p$ and $q$ being non-empty (see Sect. 1.2.4). It is *exact* if $\text{supp}(p) \cap \text{supp}(r) = \text{supp}(q) \cap \text{supp}(r)$. Such an exact conditional redescription is denoted $(p \equiv q \mid r)$.

A sketch of the procedure for mining redescriptions with the CHARM-L algorithm is provided in Algorithm 2.2. This algorithm is designed to handle cases where the data consist of a single table **D** such that each attribute belongs to its own distinct view. The algorithm first builds the closed itemset lattice for **D** using CHARM-L (line 1 in Algorithm 2.2). Each closed itemset, that is, each point in the lattice, is then considered in turn. Every pair $(J, J')$ of minimal generators of the current closed itemset $I$ generates an exact conditional redescription of the form $(X \equiv Y \mid Z)$, stored as a triple $(X, Y, Z)$ (line 6), where $X = J \setminus J'$ and $Y = J' \setminus J$ are the disjoint itemsets that form the two descriptions, while $Z = J \cap J'$ is the itemset on which the redescription is conditioned. We have $J \setminus J' \neq \emptyset$ and $J' \setminus J \neq \emptyset$ by definition of the minimal generators.

Zhao et al. (2006) later proposed BLOSOM as a generalization of CHARM for mining closed Boolean propositions beyond conjunctions, with specific closure operators also for disjunctions, as well as expressions in disjunctive normal form (DNF) and conjunctive normal form (CNF). Substituting BLOSOM for CHARM in the procedure of Algorithm 2.2 allows us to obtain conditional redescriptions involving queries that are less restricted than pure conjunctions (Ramakrishnan and Zaki 2009).

---

**Algorithm 2.2** Sketch of the procedure for mining exact conditional redescriptions with the CHARM-L algorithm

---

**Input:** A Boolean data table **D**.
**Output:** Redescriptions $\mathcal{R}$.
 1: $L \leftarrow$ closed itemset lattice built from **D** using CHARM-L
 2: $\mathcal{R} \leftarrow \emptyset$
 3: **for each** closed itemset $I$ in $L$ **do**
 4:     **for each** pair $(J, J') \in \mathcal{M}(I) \times \mathcal{M}(I)$ **do**
 5:         $X \leftarrow J \setminus J'$;   $Y \leftarrow J' \setminus J$;   $Z \leftarrow J \cap J'$
 6:         $\mathcal{R} \leftarrow \mathcal{R} \cup \{(X, Y, Z)\}$
 7:     **end for**
 8: **end for**
 9: **return** $\mathcal{R}$

---

## 2.2    Queries Based on Decision Trees and Forests

Another strategy for mining redescriptions is to use an alternating scheme. The general idea is to start with one query, find a good matching query to complete the pair, drop the first query and replace it with a better match, and continue to alternate in this way, constructing a fresh query on one or the other side until no further improvement can be achieved.

For example, in the case where we have two data tables $\mathbf{D}_1$ and $\mathbf{D}_2$, we would start with an initial query $p^{(0)}$ over $\mathbf{D}_1$ and look for a good matching query $q^{(1)}$ over $\mathbf{D}_2$. Next, we would drop $p^{(0)}$ and look for another query $p^{(2)}$ over $\mathbf{D}_1$ to form a better pair $(p^{(2)}, q^{(1)})$, and so on.

In fact, if one query of the redescription is fixed, finding an optimal query to complete the pair constitutes a binary classification task (see Sect. 1.3). The entities supporting the fixed query provide positive examples, and the remaining entities might be considered as negative examples. Thus, any feature-based classification technique could potentially make up the basis for a redescription mining algorithm, with the associated query language consisting of the possible classification criteria. However, we require interpretable queries that specify explicit constraints on the range of the values taken by the attributes. This requirement directs our choice of classification technique, precluding, for instance, the direct use of kernel support vector machines (Cortes and Vapnik 1995).

A decision tree represents a succession of tests on the value of attributes, leading to some outcome (see Aggarwal 2015, Section 10.3). The tree is called a classification tree or a regression tree, depending on whether the value to predict is a class label (the learning target is discrete) or a numerical value (the learning target is continuous), respectively. Each leaf node in the tree represents a decision, predicting a class label or a numerical value, depending on the type of tree. Each intermediate node represents a test on the value of an attribute. An example of a simple decision tree is shown in Fig. 2.2. It is a classification tree predicting a binary class label $y$ and containing tests on one Boolean attribute $a$ and two numerical attributes $b$ and $c$.

**Fig. 2.2** An example of a simple decision tree with tests on Boolean attribute $a$ and numerical attributes $b$ and $c$ (square nodes), leading to a binary classification decision (round nodes)

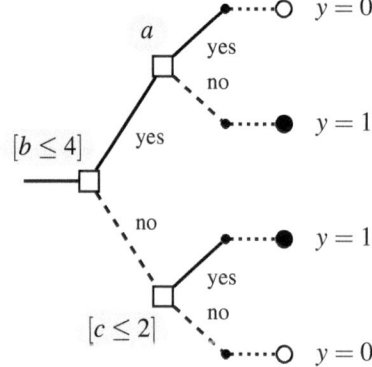

In the learning phase, decision trees are induced iteratively from the data (Breiman et al. 1984; Quinlan 1986). From a high-level perspective, the induction works as follows: For a fixed target and set of attributes and starting from a root node that contains all the entities, each attribute is evaluated in turn, computing a score that indicates how well a test on this attribute is able to discriminate between the different target values. The test that yields the best split is selected and appended to the tree as a new node, with outgoing edges representing the various possible outcomes of the test and the corresponding subsets of entities. The same splitting procedure is applied recursively and independently on each edge, considering only the associated subset of entities. This refinement process stops if one of three conditions is reached: (1) the entities in the subset all take the same value for the target, (2) no better split of the entities can otherwise be achieved, or (3) the branch has reached the maximum depth set by the user. A leaf node is then appended to the branch, containing the decision which is set to the majority label or average value among the entities in the associated subset.

Once the tree has been learnt, prediction is very straightforward. For a given entity, one simply travels through the tree, choosing which edge to follow according to the outcome of the tests in the nodes encountered along the way, starting from the root node and reaching a leaf node. The label or value prediction is indicated in that leaf node.

For a given decision tree, one can build a query by reading off the conditions of the tests along the branches of the tree. A branch from the root down to some leaf corresponds to a conjunction of predicates, and different branches can be combined with a disjunction. For example, the query

$$[b \leq 4] \wedge a$$

encodes the conditions of the topmost branch in the tree shown in Fig. 2.2, while the query

$$\big([b \leq 4] \wedge \neg a\big) \vee \big([4 < b] \wedge [c \leq 2]\big)$$

encodes the conditions leading to the positive class, through the two middle branches. Because of the way in which they are obtained, these queries follow a specific syntax, and we call them *tree-shaped queries*.

**Visualization: Tree Diagram**

Figure 2.3 shows a redescription depicted in a tree diagram. This redescription was mined by the LayeredT algorithm, which we will present shortly. The data consist of two tables; one contains records of the presence of some mammal species and the other contains temperature and precipitation, such as can be studied to find *bioclimatic niches* (see Sect. 1.1). The redescription depicted is

(continued)

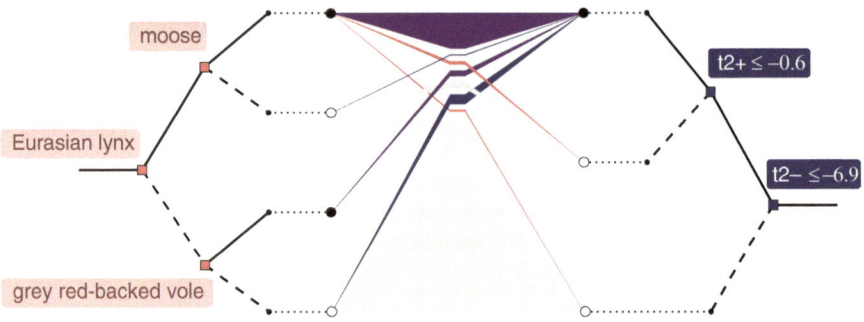

**Fig. 2.3** A redescription obtained with the `LayeredT` algorithm, depicted in a tree diagram

$$\left(\textit{Eurasian lynx} \wedge \textit{moose}\right) \vee \left(\neg\,\textit{Eurasian lynx} \wedge \textit{grey red-backed vole}\right)$$

$$\sim \; [t_2^- \leq 6.9] \wedge [t_2^+ \leq -0.6] \,.$$

On the left- and right-hand sides of the diagram are the trees built over species variables and climate variables, respectively. As in Fig. 2.2, intermediate nodes, which represent tests, are drawn as squares, while leaves, which represent decisions, are drawn as circles, with positive and negative leaves drawn as black and white circles, respectively. The two trees are joined by their leaves in the middle. The lines joining a pair of leaves of either tree represent the entities (in this case geographical regions), that simultaneously satisfy the conditions of the respective branches. For example, lines crossing from the bottommost leaf on the left to the topmost leaf on the right represent entities where neither the Eurasian lynx nor the grey red-backed vole live and where the minimum and maximum temperatures in February are below $-6.9\,°C$ and $-0.6\,°C$, respectively. The first leaf of the pair is negative and the second positive. Indeed, the entities support the climate query but not the species query. Hence these entities belong to the set $\mathcal{E}_{01}$, and the corresponding lines are drawn in dark blue. Entities belonging to $\mathcal{E}_{10}$, $\mathcal{E}_{11}$ and $\mathcal{E}_{00}$ are represented by light red, medium purple, and very light grey lines, respectively.

### 2.2.1 The `CARTwheels` Algorithm

The alternating scheme for mining redescriptions was introduced by Ramakrishnan et al. (2004), who proposed the `CARTwheels` algorithm based on the CART induction algorithm (Breiman et al. 1984). The CARTwheels algorithm was designed to handle purely Boolean data. A sketch of this algorithm is provided in Algorithm 2.3. We only outline its mechanism; more details can be found in the original publications by Ramakrishnan et al. (2004), Kumar (2007), and Ramakrishnan and Zaki (2009).

Given two Boolean data tables $\mathbf{D}_1$ and $\mathbf{D}_2$, we assume, without loss of generality, that the initial class labels are obtained from $\mathbf{D}_1$ (line 2 in Algorithm 2.3). Specifically, each Boolean attribute from $\mathbf{D}_1$ is considered as a class, and each entity is assigned to the class of the first attribute that it contains, under some arbitrary order of the attributes. The result is a one-dimensional multi-label vector, which is used as a learning target to induce a tree over $\mathbf{D}_2$ (line 3). The labels predicted by this tree will be used in turn in the next iteration to build a new tree over $\mathbf{D}_1$. This alternating process can be run for a fixed number of iterations (lines 6–12), producing a sequence of classification trees such that every other tree involves attributes from $\mathbf{D}_1$ or $\mathbf{D}_2$. Redescriptions can then be extracted from this sequence by selecting consecutive trees and generating the queries associated with some class

---

**Algorithm 2.3** Sketch of the CARTwheels algorithm

---

**Input:** Two Boolean data tables $\mathbf{D}_1$ and $\mathbf{D}_2$, similarity $\sim$, and number of iterations $\kappa$.
**Output:** Redescriptions $\mathcal{R}$.
1: $\mathcal{R} \leftarrow \emptyset$
2: $\tau_1 \leftarrow$ initialize the multi-class target with attributes from $\mathbf{D}_1$
3: $T_2 \leftarrow$ induce tree over $\mathbf{D}_2$ with target $\tau_1$
4: $\tau_2 \leftarrow$ extract classification vector from $T_2$
5: $\mathcal{T} \leftarrow (T_2)$
6: **for** iteration $k \in \{1, 2, \ldots, \kappa\}$ **do**
7:     **for** sides $(s, t) \in \{(1, 2), (2, 1)\}$ **do**
8:         $T_s \leftarrow$ induce tree over $\mathbf{D}_s$ with target $\tau_t$
9:         $\tau_s \leftarrow$ extract classification vector from $T_s$
10:        append $T_s$ to $\mathcal{T}$
11:    **end for**
12: **end for**
13: **for each** pair $(T_1, T_2)$ of consecutive trees $\mathcal{T}$ **do**
14:    **for each** class $c$ in the leaves of the trees **do**
15:        $p \leftarrow$ extract query from the branches of $T_1$ leading to class $c$
16:        $q \leftarrow$ extract query from the branches of $T_2$ leading to class $c$
17:        **if** $p \sim q$ **then**
18:            $\mathcal{R} \leftarrow \mathcal{R} \cup \{(p, q)\}$
19:        **end if**
20:    **end for**
21: **end for**
22: **return** $\mathcal{R}$

---

(lines 13–21). The procedure used to construct the trees, using the prediction from one iteration to induce the tree at the next iteration, is such that the trees are matched at the leaves and should, hence, produce good matching query pairs.

A classification tree represents a partition of the entities into classes. Each class predicted by the tree is associated with the leaves labelled with that class. In turn, these leaves are associated with the subset of entities that belong to them and with the query formed by taking the union of the branches leading to these same leaves. Each such partition can be seen as a random variable that assigns a value, in this case a class label, to each entity. For a partition of the entities into subsets given as a classification target, called the *class partition*, the aim is to induce a tree representing a partition, called the *path partition*, that matches the class partition as well as possible. That is, for a class partition corresponding to a random variable $X$, the aim is to find a path partition such that the corresponding random variable $Y$ is as informative as possible about $X$.

Entropy is typically used to measure how much information is contained in a variable. The entropy of $X$, denoted as $H(X)$, measures the amount of information needed to describe variable $X$. It is defined as

$$H(X) = -\sum_{i=1}^{n} P(x_i) \log(P(x_i)) \, ,$$

where $x_1, x_2, \ldots, x_n$ are the different values taken by variable $X$ and log denotes the binary logarithm. The entropy of $X$ conditioned on $Y$, denoted as $H(X \mid Y)$, measures the amount of information needed to describe variable $X$, given that variable $Y$ is known. The information gain $IG(X, Y)$ measures the reduction in the entropy of $X$ brought by knowing $Y$, that is, informally, how much we learn about $X$ by knowing $Y$. It is defined as $IG(X, Y) = H(X) - H(X \mid Y)$ and is equal to the mutual information of $X$ and $Y$, $I(X; Y)$. In particular, finding a classifier with the maximum information gain $IG(X, Y) = H(X)$, that is, a classifier such that the entropy $H(X \mid Y)$ reduces to zero, means that there is a one-to-one mapping between the two partitions. This will result in queries that constitute exact redescriptions, that is, redescriptions with Jaccard index equal to 1. More generally, using the information gain criterion means building a tree while trying to maximize the mutual information of $X$ and $Y$, $I(X; Y)$ and, therefore, indirectly minimizing the Jaccard distance that corresponds to the entropy distance as given by (1.14).

The goal in standard classification is to find the best match between the target labels and the prediction. However, in the context of redescription mining and especially during the first iterations, inducing sub-optimal trees can help increase the exploratory power of the algorithm. Indeed, finding a near perfect match for the current target might allow us to find a very accurate redescription, but it also means that the search will converge. Choosing sometimes the second or third best splitting test during the tree induction phase can diversify the search. In short, as Ramakrishnan et al. (2004) argue, a balance must be kept between the impurity in the classification, which drives the exploration, and the redundancy, to ensure a good coverage of the search space.

## 2.2.2 The `SplitT` and `LayeredT` Algorithms

Zinchenko et al. (2015) introduced two more tree-based redescription mining algorithms, `SplitT` and `LayeredT`, that differ in the ways in which trees are grown. Sketches of the algorithms are provided in Figs. 2.4 and 2.5, respectively. Both algorithms use a single variable to generate the initial binary targets (line 3 in Figs. 2.4 and 2.5), considering each attribute in turn. `SplitT` then grows trees on either side alternately while progressively increasing their depth, inducing a new tree from scratch in every iteration (lines 4–9 in Algorithm 2.4). `LayeredT` instead grows trees layer by layer. One layer is added to the current candidate tree by appending a new decision tree of depth one to each of its branches, each of which is learnt independently from the others (lines 11 and 16 in Algorithm 2.5). After a pair of trees has been learnt, the queries are extracted from the positive branches to form a candidate redescription. This extraction is similar in both algorithms (lines 10–14 in Algorithm 2.4 and lines 20–24 in Algorithm 2.5).

*Example 5* The three approaches for mining redescriptions with an alternating tree induction process are illustrated in Fig. 2.4. Stage *I* shows the sequence of steps going from an initial target to the obtention of a pair of trees (in bold frames). In Stage *II*, the two trees are paired and matched to extract the queries and compute the supports. The first stage is specific to each algorithm, while the second stage is the same for all three algorithms. The data in this example consist of two Boolean data tables $\mathbf{D}_1$ and $\mathbf{D}_2$ containing attributes $a$ to $d$ (in light red) and $e$ to $h$ (in dark blue), respectively, and 19 entities. To keep the example simple, the depth of the

---

**Algorithm 2.4** Sketch of the `SplitT` algorithm

**Input:** Two data tables $\mathbf{D}_1$ and $\mathbf{D}_2$, similarity $\sim$, and maximum depth $\kappa$.
**Output:** Redescriptions $\mathcal{R}$.
1: **for** sides $(s, t) \in \{(1, 2), (2, 1)\}$ **do**
2:      **for each** attribute $a_i \in \mathcal{A}_s$ **do**
3:          $\tau_s \leftarrow$ initialize the binary target with $a_i$
4:          **for each** iteration $k \in \{1, \ldots, \kappa\}$ **do**
5:              $T_t^{(k)} \leftarrow$ induce tree over $\mathbf{D}_t$ with target $\tau_s$ and depth $k$
6:              $\tau_t \leftarrow$ extract binary classification vector from $T_t^{(k)}$
7:              $T_s^{(k)} \leftarrow$ induce tree over $\mathbf{D}_s$ with target $\tau_t$ and depth $k$
8:              $\tau_s \leftarrow$ extract binary classification vector from $T_s^{(k)}$
9:          **end for**
10:          $p \leftarrow$ extract query from positive branches of $T_1^{(\kappa)}$
11:          $q \leftarrow$ extract query from positive branches of $T_2^{(\kappa)}$
12:          **if** $p \sim q$ **then**
13:              $\mathcal{R} \leftarrow \mathcal{R} \cup \{(p, q)\}$
14:          **end if**
15:      **end for**
16: **end for**
17: **return** $\mathcal{R}$

---

**Algorithm 2.5** Sketch of the LayeredT algorithm

---

**Input:** Two data tables $\mathbf{D}_1$ and $\mathbf{D}_2$, similarity $\sim$, and maximum depth $\kappa$.
**Output:** Redescriptions $\mathcal{R}$.

 1: **for** sides $(s,t) \in \{(1,2),(2,1)\}$ **do**
 2:     **for each** attribute $a_i \in \mathcal{A}_s$ **do**
 3:         $\tau_s \leftarrow$ initialize the binary target with $a_i$
 4:         $T_t^{(1,\emptyset)} \leftarrow$ induce tree over $\mathbf{D}_t$ with target $\tau_s$ and depth 1
 5:         $\tau_t \leftarrow$ extract binary classification vector from $T_t^{(1,\emptyset)}$
 6:         $T_s^{(1,\emptyset)} \leftarrow$ induce tree over $\mathbf{D}_s$ with target $\tau_t$ and depth 1
 7:         $\tau_s \leftarrow$ extract binary classification vector from $T_s^{(1,\emptyset)}$
 8:         **for each** iteration $k \in \{2,\ldots,\kappa\}$ **do**
 9:             **for each** leaf $\ell$ of $T_t^{(k-1,*)}$ **do**
10:                 $T_t^{(k,\ell)} \leftarrow$ induce tree over the subset of $\mathbf{D}_t$ contained in $\ell$
11:                                               with target $\tau_s$ and depth 1
12:             **end for**
13:             $\tau_t \leftarrow$ extract binary classification vector from the trees at level $k$, $T_t^{(k,*)}$
14:             **for each** leaf $\ell$ of $T_s^{(k-1,*)}$ **do**
15:                 $T_s^{(k,\ell)} \leftarrow$ induce tree over the subset of $\mathbf{D}_s$ contained in $\ell$
16:                                               with target $\tau_t$ and depth 1
17:             **end for**
18:             $\tau_s \leftarrow$ extract binary classification vector from the trees at level $k$, $T_s^{(k,*)}$
19:         **end for**
20:         $p \leftarrow$ extract query from positive branches of stacked trees $T_s^{(*,*)}$
21:         $q \leftarrow$ extract query from positive branches of stacked trees $T_t^{(*,*)}$
22:         **if** $p \sim q$ **then**
23:             $\mathcal{R} \leftarrow \mathcal{R} \cup \{(p,q)\}$
24:         **end if**
25:     **end for**
26: **end for**
27: **return** $\mathcal{R}$

---

trees is limited to two. In this example, the three approaches produce the same redescription, namely

$$(a \wedge c) \vee (\neg a \wedge \neg b) \sim (h \wedge e) \vee (\neg h \wedge f) \,.$$

The sizes of its support subsets are $|\mathcal{E}_{11}| = 9$, $|\mathcal{E}_{10}| = 2$, $|\mathcal{E}_{01}| = 3$, $|\mathcal{E}_{00}| = 5$, and its Jaccard index is therefore $J = 9/14$.

As in Fig. 2.2, the intermediate nodes in the trees, representing the tests, are drawn as squares. Edges corresponding to positive test outcomes (i.e. 'yes') are drawn as solid lines, while negative outcomes (i.e. 'no') are drawn as dotted lines. Subsets of entities are represented as rounded rectangles containing circles, with black and white circles representing entities assigned to the positive class and negative class, respectively. Note that the CARTwheels algorithm can involve more than two classes in general, but we only represent the case of a binary target. In the row of pictures at the top of Fig. 2.4, we see how CARTwheels alternates between the two data tables. In each iteration, a tree is induced for the given target

Stage *I*: Growing the trees

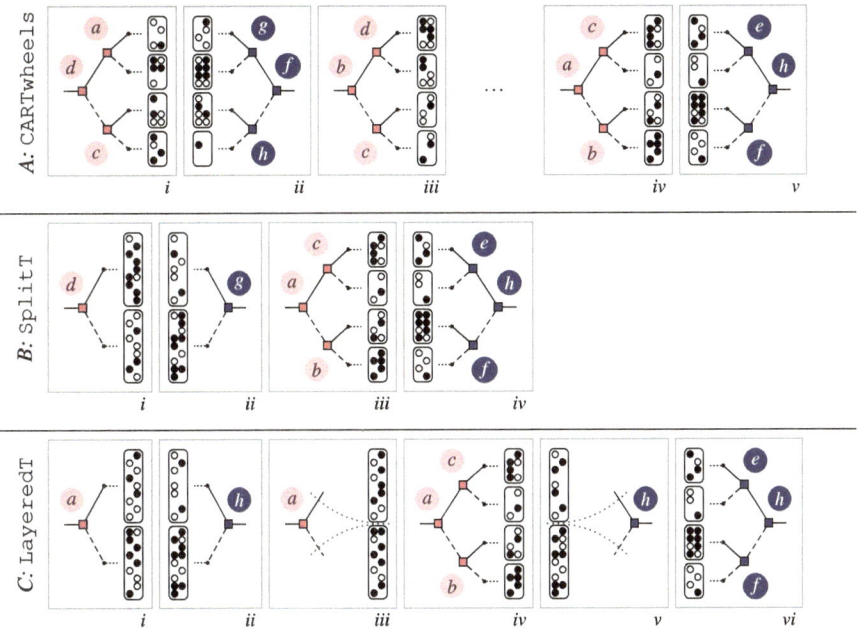

Stage *II*: Matching the trees and extracting the queries

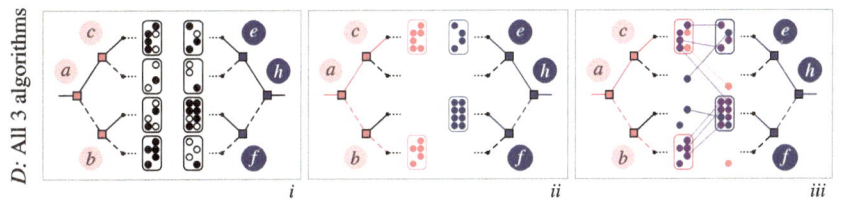

**Fig. 2.4** The CARTwheels, SplitT, and LayeredT algorithms depicted as a sequence of steps. Growing the trees (Stage *I*) is specific to each algorithm, while matching the trees (Stage *II*) is the same for all three algorithms

(step *A.i*), the entities are relabeled with the majority class and collected to form a new target, which is used in the next turn to induce a tree on the other set of attributes (step *A.ii*), and so on for a chosen number of iterations. The second row of pictures in Fig. 2.4 illustrates SplitT. In this algorithm, a tree of depth one is induced over the attributes of table $\mathbf{D}_1$ (step *B.i*) and then over the attributes of table $\mathbf{D}_2$ (step *B.ii*). Next, the target obtained from this depth-one tree is used to induce a new tree, this time of depth two, over the attributes of table $\mathbf{D}_1$ (step *B.iii*). With the resulting target, a new tree, also of depth two, is then induced over the attributes of table $\mathbf{D}_2$ (step *B.iv*). The LayeredT algorithm, depicted in the third row of pictures of Fig. 2.4, similarly builds trees of increasing depths. However, instead of building a new tree from scratch, it considers the depth-one tree obtained in the previous round and appends a new tree of depth one to either of its branches (steps *C.iii–iv* and *C.v–vi*).

Finally, in all tree algorithms, the trees are paired as shown in the last row of pictures in Fig. 2.4. For either tree, a query is obtained by combining the branches that lead to the positive class and the support of the query consists of the entities in the corresponding leaves (step *D.ii*). A redescription is formed by combining the two queries, and its support is computed by matching the entities across the two trees (step *D.iii*).

### 2.2.3  The CLUS-RM Algorithm

Mihelčić et al. (2016) present yet a different tree-based method for mining redescriptions. Their CLUS-RM algorithm, sketched in Algorithm 2.6, uses multi-target Predictive Clustering Trees (PCT).

A predictive clustering tree (Blockeel et al. 1998) is a type of decision tree that can be used to predict multiple target attributes at one time. Such a decision tree provides a succession of tests that progressively group the entities into clusters, which become more homogeneous with respect to the target attributes as one progresses down the tree. In this sense, one can see this structure as generating a hierarchical clustering of the entities.

To initialize the algorithm, Mihelčić et al. (2016) propose to use additional entities that are randomized variants of the original ones. That is, the original data set is duplicated, and the values of each attribute are shuffled among the entities in the copy. A PCT is then induced over the extended data set that contains both the original and the randomized entities, trying to discriminate between them (line 3 in Algorithm 2.6). The initial target vector is, therefore, a binary vector with a positive label for the original entities and a negative label for the randomized copies. Queries are then extracted from the resulting tree (line 4). More precisely, for each leaf and each intermediate node corresponding to a non-empty cluster of original entities (disregarding the randomized copies), the conditions encountered when travelling from the root of the tree to the node are gathered to form a conjunctive query. This procedure is applied to both data tables separately, resulting in two sets of queries $Q_1^{(0)}$ and $Q_2^{(0)}$.

For a given collection of queries $Q$, a multi-dimensional target $\tau$ can be generated as a binary matrix with one row per original entity and one column per query in $Q$, where $\tau(i,j)$ indicates whether entity $i$ belongs to the support of query $j$. A target generated from the queries obtained from one data table is used in the next iteration to induce a PCT on the original data table from the opposite side, obtaining a new collection of queries, and so on. Two such procedures are run in parallel, alternating between the sides for a fixed number of iterations (lines 7–15).

---

**Algorithm 2.6** Sketch of the CLUS-RM algorithm

---

**Input:** Two data tables $\mathbf{D}_1$ and $\mathbf{D}_2$, similarity $\sim$, and number of iterations $\kappa$.
**Output:** Redescriptions $\mathcal{R}$.
 1: $\mathcal{R} \leftarrow \emptyset$
 2: **for each** side $s \in \{1, 2\}$ **do**
 3: $\quad T_s^{(0)} \leftarrow$ induce tree over $\mathbf{D}_s$ to separate original entities from randomized copies
 4: $\quad Q_s^{(0)} \leftarrow$ extract queries from $T_s^{(0)}$
 5: **end for**
 6: append query pairs from $Q_1^{(0)} \times Q_2^{(0)}$ to $\mathcal{R}$
 7: **for** iteration $k \in \{1, \ldots, \kappa\}$ **do**
 8: $\quad$ **for** sides $(s, t) \in \{(1, 2), (2, 1)\}$ **do**
 9: $\quad\quad \tau \leftarrow$ generate multi-dimensional target from queries in $Q_s^{(k-1)}$
10: $\quad\quad T_t^{(k)} \leftarrow$ induce tree over $\mathbf{D}_t$ with target $\tau$
11: $\quad\quad Q_t^{(k)} \leftarrow$ extract queries from $T_t^{(k)}$
12: $\quad\quad$ append query pairs from $Q_s^{(k-1)} \times Q_t^{(k)}$ to $\mathcal{R}$
13: $\quad$ **end for**
14: $\quad$ append query pairs from $Q_1^{(k)} \times Q_2^{(k)}$ to $\mathcal{R}$
15: **end for**
16: combine, reduce, and prune candidate redescriptions in $\mathcal{R}$
17: **return** $\mathcal{R}$

---

In each iteration $k$ and for either side $s$, the best queries are collected from the induced tree $T_s^{(k)}$ into $Q_s^{(k)}$ (line 11). These queries are then paired with the queries in $Q_t^{(k-1)}$ collected at the previous iteration from the opposite side $t$ (line 12). The queries collected from either side at the same iteration are also combined together (line 14). These candidate redescriptions are then combined to form more complex queries potentially involving disjunctions, long conjunctions are reduced to shorter ones with the same support, and the final set of results is produced by pruning away the less accurate and redundant candidates (line 16).

To improve the quality and diversity of the obtained queries, Mihelčić et al. (2017) extended the CLUS-RM algorithm to use a forest of clustering prediction trees in addition to a single tree. Mihelčić et al. (2017) proposed a refined procedure, which they call redescription set optimization, to select the final subset of redescriptions returned as the output of the algorithm.

Finding good starting points for the alternating tree induction process is crucial for these tree-based algorithms to work. In CARTwheels, a one-dimensional multi-label vector is derived from the entire set of attributes on one side. In SplitT and LayeredT, on the other hand, the initialization target is derived from single attributes, that is, from the simplest possible queries. CLUS-RM instead uses clusters of the original entities that are formed using discriminating characteristics of the original entities when compared to randomized copies. One further option, proposed by Ramakrishnan and Zaki (2009), is to randomly partition the entities into positive and negative examples, using one or several such partitions to initialize the search, instead of actual queries.

For a fixed number of starting points and a limit on the number of alternations, the complexity of such an alternating classification scheme for building redescriptions depends primarily on the complexity of the chosen classification algorithm.

The different algorithms were presented above assuming input data in the form of two tables. More generally, the algorithms can handle arbitrary many views: the set of attributes used to build a new tree simply needs to exclude attributes participating in the tree that provides the target as well as attributes belonging to the same view as any of these attributes.

## 2.3   Growing the Queries Greedily

Finally, a third exploration strategy relies on iteratively finding the best atomic update to the current query pair. More precisely, given a pair of queries, one applies atomic operations on either query to improve the candidate redescription, until no further improvement can be achieved. Conceptually, atomic operations at hand include the addition, deletion, and editing of predicates. That is, one might add a fresh predicate to the query, remove a predicate from the query, or alter some predicate already occurring in the query, in particular, by modifying the range of the truth value assignment.

For example, if our current candidate redescription is

$$Eurasian\ lynx \vee Canada\ lynx \sim [-24.4 \leq t_3^+ \leq 3.4]$$

by adding, deleting, and editing a predicate, we might modify it to

$$Eurasian\ lynx \vee Canada\ lynx \sim [-24.4 \leq t_3^+ \leq 3.4] \wedge [5.0 \leq p_8]\ ,$$

$$Canada\ lynx \sim [-24.4 \leq t_3^+ \leq 3.4]\ ,\ or$$

$$Eurasian\ lynx \vee Canada\ lynx \sim [-24.4 \leq t_3^+ \leq 10.7]\ .$$

### 2.3.1   The ReReMi Algorithm

This strategy, restricted to the addition of predicates, that is, to extending the queries, was first introduced as the Greedy algorithm by Gallo et al. (2008). Building upon this work, Galbrun and Miettinen (2012) proposed the ReReMi algorithm, which extends the approach to handle categorical and numerical attributes along with Boolean ones and uses a beam search to keep the current top candidates at each step instead of focusing on the single best improvement. A sketch of the ReReMi algorithm is provided in Algorithm 2.7.

---

**Algorithm 2.7** Sketch of the `ReReMi` algorithm
***
**Input:** Two data tables $\mathbf{D}_1$ and $\mathbf{D}_2$, similarity $\sim$, number of initial candidates $k_p$, and beam width
    $k_i$.
**Output:** Redescriptions $\mathcal{R}$.
 1: $\mathcal{R} \leftarrow \emptyset$
 2: $\mathcal{I} \leftarrow \{k_p \text{ best initial singleton redescriptions}\}$
 3: **for** $S \in \mathcal{I}$ **do**
 4:     $\mathcal{K} \leftarrow \{S\}$
 5:     $F_1(S), F_2(S) \leftarrow$ free attributes for $S$
 6:     **if** $F_1(S) \neq \emptyset$ or $F_2(S) \neq \emptyset$ **then**
 7:        $\mathcal{L} \leftarrow \{S\}$
 8:     **end if**
 9:     **while** $\mathcal{L} \neq \emptyset$ **do**
10:        **for each** $R \in \mathcal{L}$ **do**
11:           **for** side $s \in \{1, 2\}$ and operator $\circ \in \{\vee, \wedge\}$ **do**
12:              **if** $R$ can be extended on side $s$ with operator $\circ$ and literal $l \in F_s(R)$ **then**
13:                 $\mathcal{K} \leftarrow \mathcal{K} \cup \{\text{best such extension of } R\}$
14:              **end if**
15:           **end for**
16:        **end for**
17:        $\mathcal{K} \leftarrow \{k_i \text{ best redescriptions from } \mathcal{K}, \text{ with updated free attributes}\}$
18:        $\mathcal{L} \leftarrow \{R \in \mathcal{K} : F_1(R) \neq \emptyset \text{ or } F_2(R) \neq \emptyset\}$
19:     **end while**
20:     $\mathcal{R} \leftarrow \mathcal{R} \cup \mathcal{K}$
21: **end for**
22: **return** $\mathcal{R}$

---

First, given two data tables $\mathbf{D}_1$ and $\mathbf{D}_2$, all pairs of attributes in $\mathcal{A}_1 \times \mathcal{A}_2$ are evaluated, and the best matching predicates are kept as the initial singleton redescriptions (line 2 in Algorithm 2.7). Each such initial candidate is then extended in turn, trying to append a new predicate to either query (lines 11–15) and keeping the best candidates at one step so as to be further extended in the next step (line 17). Memorization of the explored queries is used to prevent the algorithm from repeating itself. For a given candidate, this mechanism allows us to determine the attributes that will not lead to an extension already encountered previously. Such attributes are called *free attributes* (line 5). When no further extension is possible, either because there are no free attributes left, because no improvement can be achieved, or because the maximum number of iterations has been reached, the best extensions are added to the set of results (line 20).

Given a candidate redescription $R = (p, q)$ and a predicate $v$, there are four ways in which to extend $p$ with $v$: $p \wedge v$, $p \wedge \neg v$, $p \vee v$, and $p \vee \neg v$. During the extension step (line 13), barring restrictions on the types of extensions, the algorithm needs to compute the Jaccard index for four different types of extensions for each predicate. In fact, a simple observation allows us to expedite the computation of the Jaccard index for the different extensions. Indeed, to compute $J(p \wedge v, q)$, we only need to consider the entities in $\text{supp}(p)$. Since, other entities will never be in $\text{supp}(p \wedge v)$ anyway. On the other hand, entities in $\text{supp}(p)$ will be in $\text{supp}(p \vee v)$ in any case and cannot affect the value of $J(p \vee v, q)$. In order to formalize this intuition, we overload the notation defined in (1.8) for the support subsets $\mathcal{E}_{xy}$ (see Sect. 1.2.3) and denote these subsets restricted to $\text{supp}(v)$ as $\mathcal{E}_{xy}(v)$ (e.g. $\mathcal{E}_{10}(v) = \mathcal{E}_{10} \cap \text{supp}(v)$).

Using this notation, the Jaccard index of redescription $(p, q)$ can be expressed as

$$J(p, q) = \frac{|\mathcal{E}_{11}|}{|\mathcal{E}_{10}| + |\mathcal{E}_{01}| + |\mathcal{E}_{11}|} , \tag{2.1}$$

and formulas can be derived for the different extensions:

$$J(p \wedge v, q) = \frac{|\mathcal{E}_{11}(v)|}{|\mathcal{E}_{10}(v)| + |\mathcal{E}_{01}| + |\mathcal{E}_{11}|} , \tag{2.2}$$

$$J(p \wedge \neg v, q) = \frac{|\mathcal{E}_{11}| - |\mathcal{E}_{11}(v)|}{|\mathcal{E}_{10}| - |\mathcal{E}_{10}(v)| + |\mathcal{E}_{01}| + |\mathcal{E}_{11}|} , \tag{2.3}$$

$$J(p \vee v, q) = \frac{|\mathcal{E}_{11}| + |\mathcal{E}_{01}(v)|}{|\mathcal{E}_{10}| + |\mathcal{E}_{01}| + |\mathcal{E}_{11}| + |\mathcal{E}_{00}(v)|} , \text{ and} \tag{2.4}$$

$$J(p \vee \neg v, q) = \frac{|\mathcal{E}_{11}| + |\mathcal{E}_{01}| - |\mathcal{E}_{01}(v)|}{|\mathcal{E}_{10}| + |\mathcal{E}_{01}| + |\mathcal{E}_{11}| + |\mathcal{E}_{00}| - |\mathcal{E}_{00}(v)|} . \tag{2.5}$$

Notice that $\mathcal{E}_{01}$, $\mathcal{E}_{10}$, and $\mathcal{E}_{11}$ can be computed once for a given candidate redescription. Then, for each predicate we could extend it with, it is enough to perform three intersection operations to obtain $\mathcal{E}_{10}(v)$, $\mathcal{E}_{01}(v)$, and $\mathcal{E}_{11}(v)$ and be able to compute the Jaccard index of all four possible extensions. Furthermore, $|\mathcal{E}_{00}|$ and $|\mathcal{E}_{00}(v)|$ can be deduced from $\text{supp}(v)$, $\mathcal{E}$, and the other support subsets so that we do not have to consider the entities for which neither $p$ nor $q$ hold. This observation can significantly speed up the algorithm.

There is only one predicate that can be built with a Boolean attribute, but in the case of categorical and numerical attributes, in addition to evaluating the extension with either logical operator, the algorithm also needs to determine respectively the category and interval that yields the best possible predicate for that extension. This is done on-the-fly during the extension process.

For a categorical attribute, all the different categories are evaluated in turn, and the one that most improves the Jaccard index is selected. Hence, the complexity of finding the best predicate to extend a redescription with a categorical attribute grows linearly with the number of categories available. Considering the predicate constructed with categorical attribute $a$ and category $c$, (2.2) can be written as

$$J(p \wedge [a = c], q) = \frac{|\mathcal{E}_{11}([a = c])|}{|\mathcal{E}_{10}([a = c])| + |\mathcal{E}_{01}| + |\mathcal{E}_{11}|} ,$$

and similarly for the other extensions.

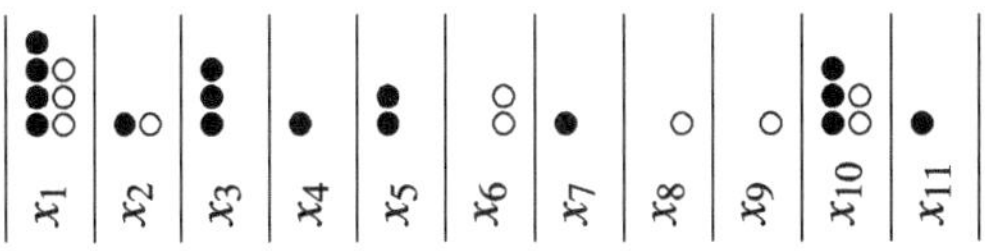

**Fig. 2.5** Example of repartition of the entities for a numerical attribute. Each bin represents a value taken by the attribute. Black circles stand for entities belonging to $\mathcal{E}_{11}$, white circles for entities from $\mathcal{E}_{10}$

For a numerical attribute, the algorithm needs to determine the lower and upper thresholds that together yield the optimal Jaccard index. Note that the two thresholds are set simultaneously, since setting one bound first and possibly the other later would prevent the search from finding some of the most specific intervals. Consider a numerical attribute $a$ together with thresholds $\lambda$ and $\rho$ in its range such that $\lambda \le \rho$, (2.2) can be written as

$$J(p \wedge [\lambda \le a \le \rho], q) = \frac{|\mathcal{E}_{11}([\lambda \le a \le \rho])|}{|\mathcal{E}_{10}([\lambda \le a \le \rho])| + |\mathcal{E}_{01}| + |\mathcal{E}_{11}|} ,$$

and similarly for the other extensions. Naively, the optimal predicate for a numerical attribute can be found by means of an exhaustive search on the acceptable thresholds. The complexity of this operation grows quadratically with the number of distinct values in the range of the attribute. At this point, the observation that not all entities can affect the support for a particular extension becomes even more useful. Indeed, only the entities in $\mathcal{E}_{11}$ and $\mathcal{E}_{10}$ can impact the Jaccard index for conjunctions and only those in $\mathcal{E}_{01}$ and $\mathcal{E}_{00}$ for disjunctions. Furthermore, only values separating entities from the two subsets, called *cut points*, need to be considered.

To illustrate this concept of cut points, assume that we are trying to set the lower threshold $\lambda$ in a conjunctive extension $p \wedge [\lambda \le a \le \rho]$ . The bins in Fig. 2.5 represent the values taken by attribute $a$, sorted in increasing order. Black circles stand for entities belonging to $\mathcal{E}_{11}$ and white circles for entities from $\mathcal{E}_{10}$. In this example, there is one entity in $\mathcal{E}_{11}$ with the value $x_4$, but none in $\mathcal{E}_{10}$ with the value $x_3$. Therefore, $x_4$ cannot be an optimal choice for $\lambda$ since choosing $x_3$ instead would always increase the accuracy. Thus, $x_3$ constitutes a cut point for the lower bound but $x_4$ does not, that is, we can consider setting $\lambda = x_3$, but setting $\lambda = x_4$ is clearly not optimal.

In summary, the running time upper bound for the greedy extension strategy is in the order of the product of the number of starting points, the maximal number of iterations, the beam width and the cost of extension tests. For a Boolean attribute, the cost of an extension test is proportional to the number of entities. It is proportional to the number of entities multiplied by the number of categories for a nominal attribute and to the squared number of entities for a numerical attribute. The data in the example of Sect. 1.1 contains approximately 55,000 entities, 48 numerical attributes (climate variables), and 4700 Boolean attributes (species). For 500 initial redescriptions, a maximum of 4 iterations, and a beam width of 4 the product above equals

$$500 \times 4 \times 4 \times (48 \times 55{,}000^2 + 4700 \times 55{,}000) \approx 10^{15} \ .$$

However, as we explained above and as argued by Galbrun and Miettinen (2012), determining the optimal extension attainable with a given numerical attribute is quadratic in the number of cut points, which is at most the number of distinct values of the attribute and usually much smaller than the number of entities. Thus, this strategy is feasible in practice.

As with tree-based approaches, it is crucial to find good starting points for the greedy extension process. One technique is to consider all possible pairs of attributes and use the $k_p$ most accurate singleton redescriptions as the starting points, with $k_p$ a parameter set by the user. In other words, we take as starting points redescriptions where the queries consist of a single predicate each. If at least one attribute in the pair is Boolean or categorical, then the best associated singleton redescription can be determined efficiently. More specifically, we apply the technique explained above to find the best extension for an empty query. Let $\emptyset$ denote an empty query, such that $\mathrm{supp}(\emptyset) = \mathcal{E}$ and when combined with a non-empty query $q$, we obtain the pair $(\emptyset, q)$ with support subsets $\mathcal{E}_{11} = \mathrm{supp}(q)$, $\mathcal{E}_{10} = \mathcal{E} \setminus \mathrm{supp}(q)$, $\mathcal{E}_{01} = \emptyset$, and $\mathcal{E}_{00} = \emptyset$. Finding the best singleton redescription for a pair of attributes $(a, b)$ where $a$ is numerical and $b$ is Boolean can be seen as finding the best extension $(\emptyset \wedge [\lambda \leq a \leq \rho], b)$. Similarly, if $b$ is categorical, we can look for how to best extend $(\emptyset, [b = c])$, considering each category $c$ in turn. On the other hand, if both attributes are numerical with many distinct values, finding the best associated singleton redescription by testing all possible combinations of thresholds for both attributes quickly becomes infeasible. In such a case, it might be necessary to first arrange the values of the attributes into buckets in order to reduce the number of thresholds that need to be tested.

The greedy extension approach was presented above assuming that the input data takes the form of two tables. However, this approach can easily be adapted to accommodate different numbers of views, by simply filtering the attributes that will be tested in the extension step. That is, for a candidate redescription $(p, q)$, attributes that belong to the same view as any attribute already appearing in query $q$ cannot be used to extend $p$ and thus do not need to be tested. This requirement can be enforced by suitably maintaining the set of available attributes with which to extend either query, that is, the free attributes $F_1$ and $F_2$.

## 2.4   A Comparative Discussion

In this chapter, we have presented the various algorithmic approaches to redescription mining. Each approach has specificities that can become advantages or drawbacks.

To begin with, queries produced with the different approaches have fairly distinctive shapes. Tree-based approaches produce queries with a specific syntax, namely, tree-shaped queries, which can be difficult to understand. This is particularly true when the query is produced by combining into a conjunction several branches of a tree that share some nodes. For example, considering a tree of depth three and combining two branches that share a test on attribute $a$ could produce the query

$$(a \wedge b \wedge \neg c) \vee (\neg a \wedge d \wedge e) \, ,$$

where attribute $a$ occurs twice, once as a positive literal and once negated. In such cases, tree diagrams, such as the one shown in Fig. 2.3, clearly help understanding and interpreting the redescription. The queries obtained with the greedy extension approach also assume a particular shape, since they result from iteratively appending predicates. They can also become difficult to parse, especially if the operators are used in alternation during the extension process, producing, for example, a query of the form

$$\big((a \wedge b) \vee c\big) \wedge d \, .$$

With tree-based approaches, it is possible to obtain a pair of queries that both involve disjunctions, if they both combine several tree branches. By default, greedy extension approaches allow a rather flexible use of the disjunction operator, and nothing prevents building a redescription with disjunctions appearing in both queries. However, redescriptions involving disjunctions in both queries tend to be fairly difficult to interpret, since it is impossible to know which conditions co-occur most often without looking closer at the support, for instance, by visualizing the redescription as a tree diagram (see Fig. 2.3) or in a parallel coordinates plot (see Fig. 1.2). A balance needs to be found between the types of queries that the algorithms can build and what is interpretable and interesting, depending on the context. It can be useful to forbid conjunctions from appearing in both queries simultaneously.

Successive extensions during the greedy process tend to have diminishing returns. In other words, the first attributes are typically responsible for the bulk of the support, while attributes added later on often merely correct a few misclassified entities. To circumvent the issue, one might define a minimum contribution requirement. That is, one might set a threshold and consider the addition of an attribute to constitute an acceptable extension only if it modifies at least that many entities in the support of the redescription. The same issue with tree-based approaches can be addressed similarly by setting a threshold on the minimum number of entities per leaf. However, this solution is rather crude and has its shortcomings, starting with the choice of the threshold value, which might be rather arbitrary. More generally, it is not easy to find *balanced* disjunctions and conjunctions, that is, expressions where different attributes contribute equivalent proportions of the support. Queries obtained with tree-based approaches are somewhat less flexible than those obtained with the greedy extension approach but also seem to be less subject to overfitting, according to the experiment of Zinchenko et al. (2015), who compared the ability of redescriptions obtained with the `SplitT`, `LayeredT`, and `ReReMi` algorithms to generalize to unseen data.

Finding good starting points is critical for the initialization of greedy extension algorithms and tree-based algorithms, but can be difficult and computationally expensive, especially with purely numerical data. In that sense, the fact that mine-and-pair algorithms do not require starting points can be seen as an advantage of these approaches.

Distributed computing provides rather straightforward means to scale redescription mining algorithms up, allowing them to handle significantly larger data sets than what would otherwise be feasible. In the case of greedy extension approaches, several initial candidates can be extended in parallel. In the case of tree-based approaches, the alternating tree-induction process can similarly be run in parallel with different initial targets. On the other hand, parallel pattern mining algorithms (see, for instance, Négrevergne et al. 2014) might offer opportunities for mine-and-pair approaches to benefit from the spread of multi-core architectures and distributed computing power.

## 2.5  Handling Missing Values

When the values are missing for some entries in the input data, the question arises as to how to handle the entities for which part of the information is missing. For an entity that contains missing entries, a given query might evaluate to true, to false, or its status might remain undetermined.

For example, imagine that we are trying to classify an entity according to the decision tree shown in Fig. 2.2. To do so, we need to travel through the tree according to the outcome of the successive tests, from the root down to a leaf. If the value of attribute $b$ is missing for the entity, we cannot choose what branch to follow after the first test, since the outcome of the test is undetermined. But considering the query built by combining the two middle branches of the decision tree:

$$([b \leq 4] \wedge \neg a) \vee ([4 < b] \wedge [c \leq 2]),$$

we might be able to determine its status for a given entity if the value of attributes $a$ and $c$ are available. Indeed, if $a$ is false and the value of $c$ is below 2 for the entity, it will satisfy the query regardless of the value of $b$. And vice-versa, if $a$ is true and the value of $c$ is strictly above 2, the entity will not satisfy the query, whatever value $b$ might take. On the other hand, if $a$ is true while the value of $c$ is below 2, if $a$ is false while the value of $c$ is strictly above 2, or if the values of $a$ or $c$ are missing, then we cannot determine whether or not the query is satisfied. All we can claim in these cases is that there exists an assignment of values to the missing entries so that the query will be satisfied.

Such a claim assumes that the attributes are independent. Furthermore, it is straightforward to determine satisfiability when attributes are required to occur at most once in a given query. When queries are in disjunctive normal form (DNF), as is the case of tree-shaped queries, it is sufficient that one of the conjunctions be satisfiable, and the satisfiability of each one can be checked separately. Hence, satisfiability can be determined easily in this case, too. When considering an unconstrained query language, however, determining whether there exists an assignment of values to the missing entries such that the query holds actually reduces to the standard Boolean satisfiability problem (SAT), an NP-complete problem (Garey and Johnson 2002).

Having evaluated the status of the two queries $p$ and $q$ for all the entities, we can define subsets of entities as in (1.8) (see Sect. 1.2.3), but extended with the undetermined status. That is, we let $\mathcal{E}_{1?}$ be the set of entities for which $p$ holds true, but the status of $q$ cannot be determined, and similarly for $\mathcal{E}_{?1}$, $\mathcal{E}_{0?}$, $\mathcal{E}_{?0}$, and $\mathcal{E}_{??}$.

Then, one can consider different ways to compute the Jaccard index depending on what values the missing entries are assumed to take, if any. Making no assumption at all and simply leaving out entities with undetermined status in either of the queries, Galbrun and Miettinen (2012) suggest to use the standard Jaccard index from (1.4), called *rejective Jaccard index* ($J_R$) in this context. Galbrun and Miettinen (2012) also propose two additional measures that represent the lower and upper bounds on the Jaccard index. That is, the measures evaluate the Jaccard index under the least favorable and most favorable assignments of values to the missing entries and are called the *optimistic Jaccard index* ($J_O$) and the *pessimistic Jaccard index* ($J_P$). Mihelčić et al. (2016) introduce an additional measure, which relies on the fact that entities in $\mathcal{E}_{11}$, $\mathcal{E}_{10}$, and $\mathcal{E}_{1?}$ are known to be in the support of $p$, while entities in $\mathcal{E}_{11}$, $\mathcal{E}_{01}$, and $\mathcal{E}_{?1}$ are known to be in the support of $q$. They rewrite the formula from (1.4), setting $\mathrm{supp}(p) = \mathcal{E}_{11} \cup \mathcal{E}_{10} \cup \mathcal{E}_{1?}$ and $\mathrm{supp}(q) = \mathcal{E}_{11} \cup \mathcal{E}_{01} \cup \mathcal{E}_{?1}$. In fact, this corresponds to assuming that the missing entries will take values so as to make the queries evaluate to false for entities that contain missing information. We call this variant the *negative Jaccard index* ($J_F$). Finally, as a counterpart to this latter variant, we obtain the *positive Jaccard index* ($J_T$) by assuming that the missing entries will take values so as to make the queries evaluate to true for entities that contain missing information.

In summary, the options for computing the Jaccard similarity index in the presence of missing values include the following variants[1]:

$$J_R(p,q) = \frac{|\mathcal{E}_{11}|}{|\mathcal{E}_{10}| + |\mathcal{E}_{01}| + |\mathcal{E}_{11}|} \,, \tag{2.6}$$

$$J_P(p,q) = \frac{|\mathcal{E}_{11}|}{|\mathcal{E}_{10}| + |\mathcal{E}_{01}| + |\mathcal{E}_{11}| + |\mathcal{E}_{1?}| + |\mathcal{E}_{?1}| + |\mathcal{E}_{0?}| + |\mathcal{E}_{?0}| + |\mathcal{E}_{??}|} \,, \tag{2.7}$$

---

[1] The equation for the pessimistic Jaccard index presented by Galbrun and Miettinen (2012, Equation 5.7) is erroneous, as it misses two summands from the denominator.

$$J_O(p,q) = \frac{|\mathcal{E}_{11}| + |\mathcal{E}_{1?}| + |\mathcal{E}_{?1}| + |\mathcal{E}_{??}|}{|\mathcal{E}_{10}| + |\mathcal{E}_{01}| + |\mathcal{E}_{11}| + |\mathcal{E}_{1?}| + |\mathcal{E}_{?1}| + |\mathcal{E}_{??}|} \; , \tag{2.8}$$

$$J_F(p,q) = \frac{|\mathcal{E}_{11}|}{|\mathcal{E}_{10}| + |\mathcal{E}_{01}| + |\mathcal{E}_{11}| + |\mathcal{E}_{1?}| + |\mathcal{E}_{?1}|} \; , \text{ and} \tag{2.9}$$

$$J_T(p,q) = \frac{|\mathcal{E}_{11}| + |\mathcal{E}_{1?}| + |\mathcal{E}_{?1}| + |\mathcal{E}_{??}|}{|\mathcal{E}_{10}| + |\mathcal{E}_{01}| + |\mathcal{E}_{11}| + |\mathcal{E}_{1?}| + |\mathcal{E}_{?1}| + |\mathcal{E}_{0?}| + |\mathcal{E}_{?0}| + |\mathcal{E}_{??}|} \; . \tag{2.10}$$

Among the algorithms mentioned in this chapter, both ReReMi (Galbrun and Miettinen 2012) and CLUS-RM (Mihelčić et al. 2016) are able to handle missing entries in the input data. CART and PCT algorithms that accept missing values in the input need to handle them during the learning phase, especially when evaluating the split obtained with a given attribute. Some algorithms might naively consider the test to yield a negative outcome when encountering a missing value. Other algorithms instead consider missing entries as taking a distinct value, and so they add a branch for the outcome corresponding to the value missing. Yet another approach consists in distributing the entities for which the value is missing between the different branches at random, while following the same proportions as observed among the other entities. In the greedy extension approach, the algorithm simply considers nine subsets of entities instead of the standard four and substitutes one of the variants above for the standard Jaccard index.

In summary, the choice of an algorithm for mining redescriptions must take into account, in particular, the restrictions that might be imposed on the query language, the type of variables involved, and the size of the data. It can be useful to try different algorithms, but comparing the resulting sets of redescriptions is typically not straightforward.

# References

Aggarwal CC (2015) Data Mining: The Textbook. Springer, Cham, https://doi.org/10.1007/978-3-319-14142-8

Agrawal R, Srikant R (1994) Fast algorithms for mining association rules in large databases. In: Proceedings of 20th International Conference on Very Large Data Bases (VLDB'94), pp 487–499

Blockeel H, De Raedt L, Ramon J (1998) Top-down induction of clustering trees. In: Proceedings of the 15th International Conference on Machine Learning (ICML'98), pp 55–63

Breiman L, Friedman J, Stone CJ, Olshen RA (1984) Classification and regression trees. CRC press, Boca Raton, FL

Cortes C, Vapnik V (1995) Support-vector networks. Mach Learn 20(3):273–297, https://doi.org/10.1007/BF00994018

Galbrun E, Miettinen P (2012) From black and white to full color: Extending redescription mining outside the Boolean world. Stat Anal Data Min 5(4):284–303, https://doi.org/10.1002/sam.11145

Gallo A, Miettinen P, Mannila H (2008) Finding subgroups having several descriptions: Algorithms for redescription mining. In: Proceedings of the 8th SIAM International Conference on Data Mining (SDM'08), pp 334–345, https://doi.org/10.1137/1.9781611972788.30

Ganter B, Wille R (1999) Formal Concept Analysis: Mathematical Foundations. Springer, Berlin, https://doi.org/10.1007/978-3-642-59830-2

Garey MR, Johnson DS (2002) Computers and intractability. A guide to the theory of NP-completeness, vol 29. W. H. Freeman and Co., San Francisco, CA

Kumar D (2007) Redescription mining: Algorithms and applications in bioinformatics. PhD thesis, Department of Computer Science, Virginia Polytechnic Institute and State University

Mannila H, Toivonen H, Verkamo AI (1994) Efficient algorithms for discovering association rules. In: Proceedings of the 1994 AAAI Workshop on Knowledge Discovery in Databases (KDD'94), pp 181–192

Mihelčić M, Džeroski S, Lavrač N, Šmuc T (2017) A framework for redescription set construction. Expert Syst Appl 68:196–215, https://doi.org/10.1016/j.eswa.2016.10.012

Mihelčić M, Džeroski S, Lavrač N, Šmuc T (2016) Redescription mining with multi-target predictive clustering trees. In: Proceedings of the 4th International Workshop on the New Frontiers in Mining Complex Patterns (NFMCP'15), pp 125–143, https://doi.org/10.1007/978-3-319-39315-5_9

Mihelčić M, Džeroski S, Lavrač N, Šmuc T (2017) Redescription mining augmented with random forest of multi-target predictive clustering trees. J of Intell Inf Syst pp 1–34, https://doi.org/10.1007/s10844-017-0448-5

Négrevergne B, Termier A, Rousset M, Méhaut J (2014) Para miner: A generic pattern mining algorithm for multi-core architectures. Data Min Knowl Disc 28(3):593–633, https://doi.org/10.1007/s10618-013-0313-2

Quinlan J (1986) Induction of decision trees. Mach Learn 1(1):81–106, https://doi.org/10.1023/A:1022643204877

Ramakrishnan N, Zaki MJ (2009) Redescription mining and applications in bioinformatics. In: Chen J, Lonardi S (eds) Biological Data Mining, Chapman and Hall/CRC, Boca Raton, FL

Ramakrishnan N, Kumar D, Mishra B, Potts M, Helm RF (2004) Turning CARTwheels: An alternating algorithm for mining redescriptions. In: Proceedings of the 10th ACM SIGKDD International Conference on Knowledge Discovery and Data Mining (KDD'04), pp 266–275, https://doi.org/10.1145/1014052.1014083

Zaki MJ, Hsiao CJ (2005) Efficient algorithms for mining closed itemsets and their lattice structure. IEEE Trans Knowl Data En 17(4):462–478, https://doi.org/10.1109/TKDE.2005.60

Zaki MJ, Ramakrishnan N (2005) Reasoning about sets using redescription mining. In: Proceedings of the 11th ACM SIGKDD International Conference on Knowledge Discovery and Data Mining (KDD'05), pp 364–373, https://doi.org/10.1145/1081870.1081912

Zhao L, Zaki MJ, Ramakrishnan N (2006) BLOSOM: A framework for mining arbitrary Boolean expressions. In: Proceedings of the 12th ACM SIGKDD International Conference on Knowledge Discovery and Data Mining (KDD'06), pp 827–832, https://doi.org/10.1145/1150402.1150511

Zinchenko T, Galbrun E, Miettinen P (2015) Mining predictive redescriptions with trees. In: IEEE International Conference on Data Mining Workshops, pp 1672–1675, https://doi.org/10.1109/ICDMW.2015.123

# Chapter 3
# Applications, Variants, and Extensions of Redescription Mining

Having formally defined the task of mining redescriptions in Chap. 1, and having presented different algorithmic techniques to carry it out in Chap. 2, we now look at it from different perspectives; specifically, we consider applications, variants, and extensions of the task.

First, we outline different applications of redescription mining, as examples of how the method can be used in various domains. Next, we present two problem variants, namely relational redescription mining and storytelling. The former aims at finding alternative descriptions for groups of objects in a relational data set, while the goal in the latter is to build a sequence of related queries in order to establish a connection between two given queries. Finally, we point out extensions of the task that constitute possible directions for future research. In particular, we discuss how redescription mining could be augmented with richer query languages and consider going beyond pairs of queries to multiple descriptions.

## 3.1 Applications of Redescription Mining

In this section, we present some applications of redescription mining. The applications come from a diverse set of domains, highlighting the versatility of redescription mining. Because our purpose in this section is to provide ideas as to how redescription mining can be applied on actual use cases, we will not explain each application in detail, instead referring the interested readers to the original publications.

© The Author(s) 2017
E. Galbrun, P. Miettinen, *Redescription Mining*, SpringerBriefs
in Computer Science, https://doi.org/10.1007/978-3-319-72889-6_3

## 3.1.1  In Biology

In the field of bioinformatics, Kumar (2007) used the CARTwheels algorithm
(Ramakrishnan et al. 2004; see also Sect. 2.2) to perform a cross-taxonomic and
cross-genomic analysis of gene ontologies, which was also discussed by Ramakrishnan and Zaki (2009). A gene ontology is a unified controlled vocabulary that
describes genes and their products. It is structured as a directed acyclic graph
of concepts. The Gene Ontology (GO)[1] contains ontologies relating to biological
processes (denoted as BIO), cellular components (CEL), and molecular functions
of genes (MOL), all gathered into one database and accompanied by a set of tools
to browse this database and use it in experiments. The terms from these ontologies
are used to annotate genes, that is, they are added to the genes as a means to explain
their role, function, etc. Some of the terms in this database are used to annotate genes
of different species. The Gene Ontology is a major standardization and annotation
initiative, which is part of the more general Open Biomedical Ontologies (OBO)
classification effort in bioinformatics.

The goals of applying redescription mining include (1) the functional enrichment
of unclassified genes, (2) the analysis of structural (in)consistencies among ontologies, and (3) communication on the similarities and differences across ontologies.

Kumar (2007) contributes to this third goal by considering the genome of
a species of roundworm (*Caenorhabditis elegans*) and mining cross-taxonomy
redescriptions, that is, identifying associations across different ontologies. For
instance, he provides an exact redescription supported by 3 genes involving terms
from GO BIO and GO CEL, on the one hand, and a term from GO MOL, on the
other hand:

$$GO : 0006415 \ (BIO \ translational \ termination) \wedge GO : 0005737 \ (CEL \ cytoplasm)$$

$$\equiv \ GO : 0003747 \ (MOL \ translation \ release \ factor \ activity) \ .$$

Another redescription he provides also has a Jaccard similarity of 1 and involves
terms from all three ontologies:

$$GO : 0003924 \ (MOL \ GT \ Pase \ activity)$$

$$\wedge \ GO : 0015630 \ (CEL \ microtubule \ cytoskeleton)$$

$$\equiv \ GO : 0046785 \ (BIO \ microtubule \ polymerization) \ .$$

This redescription is supported by 45 genes of the human genome, 30 genes of
the mouse genome, and 16 genes of the worm genome and is, therefore, cross-
taxonomic and cross-genomic at the same time.

---

[1]http://www.geneontology.org/. Accessed 25 Oct 2017.

The question of relating the genotype and the phenotype of organisms is of major importance in bioinformatics and intuitively lends itself to an analysis through redescription mining. This kind of question arises, for instance, when considering budding yeast (*Saccharomyces cerevisiae*) and looking for associations between the gene expression levels measured in different stress experiments, such as desiccation and heat shock, and the GO terms with which the genes are annotated (Ramakrishnan et al. 2004; Kumar 2007; Ramakrishnan and Zaki 2009). Such a study can help understand stress response mechanisms and relate the reaction to stress—a behavioural characteristic which is part of this organism's phenotype— and the organism's genotype.

In a laboratory experiment reported by Singh et al. (2005), yeast colonies were left to dry for 42 h, then rehydrated. Samples were taken at timestamps $T_0 = 0$ h, $T_1 = 18$ h, and $T_2 = 42$ h during the desiccation as well as $T_3 = 15$ min, $T_4 = 45$ min, $T_5 = 90$ min, and $T_6 = 6$ h during the rehydration, and gene expression levels (EL) were measured for these various samples using microarray tests. The expression levels from this desiccation/rehydration experiment were supplemented with expression levels obtained in related experiments and reported in the literature (such as a heat shock experiment) as well as with terms from the Gene Ontology.

The two redescriptions below, reported by Kumar (2007, Chapter 4), are examples of the results obtained in this study, using the CARTwheels algorithm. The first redescription is an exact redescription ($J = 1$) supported by a single gene (SIP18). It relates terms from the GO CEL ontology to expression changes during desiccation:

$$GO : 0005625 \text{ (CEL soluble fraction)} \wedge GO : 0005634 \text{ (CEL nucleus)}$$
$$\equiv [6 \leq EL@T_4 \text{ vs. } EL@T_0 \text{ desiccation}] \wedge [EL@T_4 \text{ vs. } EL@T_2 \text{ desiccation} \leq 1] .$$

The second redescription relates a particular expression level during a heat shock to a change in expression during desiccation:

$$[EL@30 \text{ min } heat\ shock \leq 1] \sim [-5 \leq EL@T_1 \text{ vs. } EL@T_0 \text{ desiccation} < -1]$$

It has a Jaccard similarity of 0.71 and is supported by 32 genes.

To better understand the phenomenon highlighted by the second redescription above and uncovering the corresponding pathway, Ramakrishnan and Zaki (2009) consider the genes in its support that do not encode for ribosomal activity and look closer into the relationships that exist between the remaining genes, based on their expression profiles and on knowledge from the literature. This way, they can reconstruct a network of gene interactions related to methyl group transfer and sulfur metabolism.

As part of a study of the links between the genotypes and phenotypes of various strains of *Staphylococcus aureus* bacteria, Gaidar (2015) applied redescription mining to relate gene expression and cell binding data for this bacteria. More precisely, she considered a collection of 29 methicillin resistant and 29 methicillin sensitive strains of *Staphylococcus aureus* collected from nasal colonisation of incoming hospital patients. Gene expression data collected through micro-array experiments for these 58 strains constituted one data table. The other data table contained the results of a whole-blood experiment, bringing the different strains in contact with fresh human blood and measuring the time needed for the bacteria to attach to particular white cell populations (granulocytes, monocytes, and lymphocytes). Thus, the attributes from the first data table characterise the genotype of the bacteria strains while those from the second data table relate to their phenotype.

Redescriptions were mined from this pair of data tables using the ReReMi algorithm (Galbrun and Miettinen 2012; see also Sect. 2.3). For instance, the most accurate redescription reported by Gaidar (2015) states that strains showing activation for genes specific to capsule type 5 production (*capH5*) are roughly the same strains for which the Mean Fluorescent Intensity (MFI) over granulocytes at the 5 min time point was lower than 886.73 arbitrary units of fluorescent intensity[2]:

$$capH5 \sim [Granulocytes\ MFI\ @5\min \leq 886.73]\ .$$

This redescription has a support of 32 strains and a Jaccard similarity of 0.889.

The obtained redescriptions highlight associations between the genotypical profile of the strains and a particular aspect of their phenotypical profile, namely their reaction in a whole-blood experiment. These results generally agree with previous experimental findings (e.g. Watts et al. 2005) and accepted knowledge in the scientific community.

Along a different line of work, Mihelčić et al. (2017) used redescription mining to relate clinical and biological characteristics of cognitively impaired patients, with the aim of improving the early diagnosis of Alzheimer's disease. They applied the CLUS-RM algorithm (Mihelčić et al. 2016; see also Sect. 2.2) on data provided by the Alzheimer's disease Neuroimaging Initiative (ADNI).[3] In this study, one data table consists of biological attributes derived from neuroimaging, from blood tests, and from genetic markers, for instance, while the other data table contains clinical attributes that record patients' answers to several questionnaires, records of observations by physicians, and results of cognition tests. The results obtained largely confirmed the findings of previous studies. In addition, they highlighted some additional biological factors whose relationships with the disease require further investigation, such as the pregnancy-associated plasma protein-A (PAPP-A), which they found to be highly associated with cognitive impairment in Alzheimer's disease.

---

[2]Units of fluorescent intensity depend on the measuring device and the procedure used, hence they are called arbitrary units.

[3]http://www.adni-info.org/. Accessed 25 Oct 2017.

## 3.1.2 In Ecology

One application of redescription mining in the field of ecology is in finding bioclimatic niches. A *bioclimatic niche*[4] (or bioclimatic envelope) explains the distribution of species based on bioclimatic properties, in contrast to other types of niches that might consider other explanatory parameters such as predation and competition relationships. Bioclimatic niches can be used to predict the ability of species to survive the effects of climate change (Pearson and Dawson 2003). Indeed, if the climate model predicts that the niche will disappear, the species is probably in danger of extinction.

The idea of finding a climate that correlates with a species was proposed in the twenties (see Pearson and Dawson 2003). More recently, Thuiller et al. (2009) presented the BIOMOD ensemble forecasting platform, which can use different algorithms for finding the niches and has become a rather popular tool for bioclimatic niche modelling. Foremost among methods used to learn niches are those based on the maximum entropy principle (Phillips et al. 2006). The choice of features to consider as input attributes—whether to consider temperature and rainfall, squares and products thereof (as variance and co-variance), land use, altitude, etc.—can have a significant effect on the obtained niches.

Galbrun and Miettinen (2012) presented an example for the application of redescription mining to the task of niche modelling. We can see the query over the species and the query over the climatic attributes as defining the *observed niche* and the *simulated niche*, respectively. Our first introductory example in Sect. 1.1 comes from this application domain, defining the bioclimatic niche of the Eurasian and Canada lynxes.

Mining redescriptions over species and bio-ecological attributes can allow scientists to identify niches that can involve multiple species. In addition, while the standard methods require manually selecting the species whose niche we are interested in, when using redescriptions, the species are selected automatically during the mining process. Besides, the obtained redescriptions are interpretable and provide sharp limits on climate conditions instead of the weights returned by regression models or the correlations detected by statistical procedures such as PCA, for instance.

When the redescriptions are intended to be used for predictions, it is important to test how well they generalize. Cross-validation is the most common test for this purpose, but splitting the data into training and testing sets is not trivial in this application. Indeed, the simplest approach is to use uniformly random samples (Phillips et al. 2006), but this presents a risk of missing small niches completely. Instead, in order to account for the North–South trends in the climate, Zinchenko et al. (2015) proposed sampling entities, that is, geographic sites, along North–South stripes. Such a sampling approach ensures better coverage of the various climates encountered along different latitudes, and hence, improves the representativity of the subsets.

---

[4]The term is used in its Grinnellian sense, see Soberón and Nakamura (2009).

Instead of modelling the distributions of species directly, one might look at the distributions of functional traits of species. Depending on the aim of the study, the traits of interest vary, including physiological, morphological, or anatomical features such as the body size, weight and shape, the diet type, the growth rate, and so on. This might allow the analyst to find associations that generalize better across space (to similar species on different continents) and across time (to extinct and fossil species that share similar traits). Galbrun et al. (2017) consider dental traits of large plant eating mammals and bioclimatic variables (derived from temperature and precipitation records) from around the globe, looking for associations between teeth features and climate. Indeed, the teeth of plant-eating mammals constitute an interface between the animal and the plant food available in its environment. Hence, teeth are expected to match the types of plant food present in the environment, and dental traits are thus expected to carry a signal of environmental conditions. In this study, three global zones are identified, namely a boreal-temperate moist zone, a tropical moist zone, and a tropical-subtropical dry zone, each associated to particular teeth characteristics and a specific climate.

For instance, the following redescription characterizes sites near the equator in Africa, South America, and Asia—sites that correspond to the tropical moist zone—in terms of the distribution of dental traits among the species that inhabit those sites, on the one hand, and of climatic variables, on the other hand.

$$([0.846 \leq Hyp1] \wedge [OL \leq 0.4]) \vee [0.033 \leq OT \leq 0.138] \wedge [Hyp3 \leq 0.348]$$

$$\sim [67 \leq TIso] \wedge [17.7 \leq T^+WarmM \leq 35.8] .$$

The first query involves traits related to the durability of the teeth through their shape ($Hyp1$ and $Hyp3$) as well as related to the presence of cutting structures ($OL$) and to occlusion properties ($OT$). The second query selects sites with a hot climate ($T^+WarmM$) and low temperature seasonality ($TIso$).

### 3.1.3  In Social and Political Sciences and in Economics

van Leeuwen and Galbrun (2015) and Galbrun and Miettinen (2016) applied redescription mining to political opinion poll data. In particular, they both used voting advice application data from Finnish parliamentary elections. An online *voting advice application* (VAA for short) is an online platform that aims to help the voters choose which party or candidate to vote for. They can do this by presenting a set of questions to the candidates and recording their answers. The same questions are then asked from the voter using the application, and finally, the application shows how well the voter's answers match with various candidates' answers. The questions are different in every election and might vary from one voting district to the next, as the developers of the VAA try to devise questions that are important to the voters in the election at hand and are divisive enough to allow for meaningful recommendations.

The experiments by van Leeuwen and Galbrun (2015) and Galbrun and Miettinen (2016) used the answers of the candidates together with their socio-economical background information (e.g. age, education level, party membership), which was released as open data by the makers of the VAAs. They both used data from the 2011 Finnish parliamentary elections, and Galbrun and Miettinen (2016) also used data from the 2015 parliamentary elections.

This kind of VAA data presents a natural setting for redescription mining: we have one data table that contains the socio-economical background and another one that contains the answers. This allows us to analyse whether the socio-economical background explains the candidates' opinions on certain topics; for example, Galbrun and Miettinen (2016) found the following redescription from the 2011 election data, describing the candidates who were against more nuclear power as being of a certain age, having a high level of education, or not being a member of the parliament at the time:

$$[51 \leq \text{Age} \leq 58] \vee [7 \leq \text{EduLvl}] \vee \neg \text{MP} \sim \neg \text{Q3.NuclearPow} .$$

The above redescription has Jaccard similarity of 0.66 with support of 366 (out of 675 candidates in the data).

**Visualization: 2D Embeddings**

Embedding high-dimensional data in two or three dimensions for visualization is a common and well-studied problem. There are many different methods for the embedding, usually aiming to preserve different features of the original data. Notable methods include principal component analysis, multidimensional scaling, self-organizing maps (Kohonen 1989), Isomap embedding (Tenenbaum et al. 2000), and locally-linear embeddings (Roweis and Saul 2000), to name but a few. These and other embeddings can also be used to visualize the entities and their relation to the support of a redescription. Figure 3.1 shows an Isomap embedding of the candidates in the 2011 Finnish parliamentary election. The colours in Fig. 3.1 are based on the above redescription: the light red squares correspond to candidates whose socio-economical background match the description but who do support more nuclear power, the medium purple squares correspond to candidates to whom both of the descriptions apply (i.e. they are in the support of the redescription), the dark blue squares correspond to candidates who do not support more nuclear power but whose socio-economical background are not as described, and the very light grey squares correspond to the candidates to whom neither of the descriptions apply.

Different embeddings will reveal different structures from the data and can be useful when analysing different redescriptions. The Isomap embedding in Fig. 3.1, for example, embeds most of the light red markers to the right and most of the medium purple markers to the left part of the plot. The few dark

(continued)

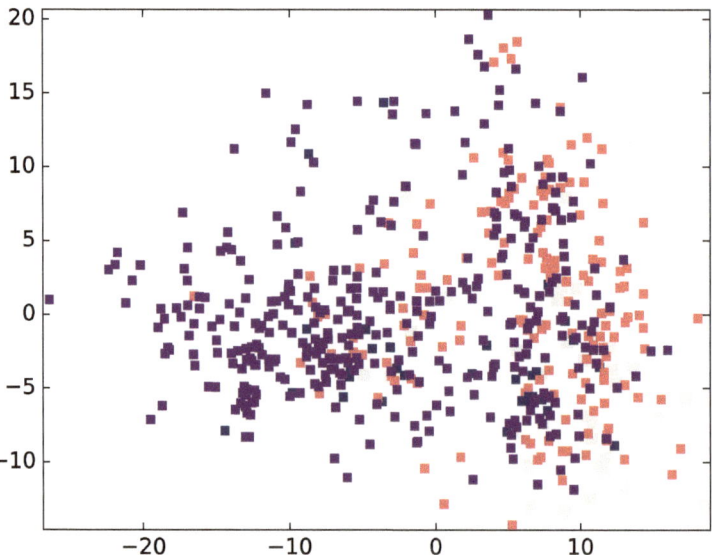

**Fig. 3.1** The data for the redescription about age, education level, political experience, and opinions towards nuclear power projected to 2D using Isomap embedding on the Boolean and categorical attributes. Light red corresponds to candidates with correct traits but who support nuclear power, medium purple represents candidates to whom both descriptions apply, dark blue is for candidates who do not support nuclear power, but whose socio-economical traits are different from the described, and very light grey is the candidates to whom neither of the descriptions apply. Candidates with missing answers are not shown

blue markers are mixed among the medium purple markers; this indicates that the socio-economical background does not really affect the embedding. However, the candidates who do support building more nuclear power are generally on the right, while the candidates who are against it are more on the left side of the plot. No clear division can be seen, though, indicating that this redescription captures some *local* pattern that is not observable at the scale of the entire data set.

Another natural question for this kind of data is how the answers in one year relate to the answers in another year. Galbrun and Miettinen (2016) compared the answers of the candidates who ran for a seat in both 2011 and 2015 elections. The following redescription is an interesting example of their findings:

$$[\text{Q17:NATO} \neq \text{YesNotSoon}] \wedge \neg \text{Q31:GvtPrt:RKP}$$

$$\wedge\, [\text{Q27:MunOutsource} \neq \text{IncButChoose}]$$

$$\sim [\text{Q137.NATO.is.good} \leq -1]\,.$$

This redescription indicates that the candidates who in 2011 did not choose the option 'yes, but not too soon' to the question 'Should Finland apply for a membership in NATO?', did not want the Swedish People's Party in the government, and did not want to increase the outsourcing of municipal services are approximately the same candidates who in 2015 disagreed with the claim 'Joining NATO would improve Finland's national security.' The redescription has a Jaccard similarity of 0.78 and a support of 368 (out of 675 candidates in the data). While the support of Finland's NATO membership does not strictly follow party lines, the redescription indicates that resistance towards membership is stronger among the politicians who share traditional left-wing opinions.

Many natural applications of redescription mining involve socio-economical data. For another example, Mihelčić et al. (2017) studied the import and export data of countries together with general data from those countries (e.g. demographical data, health-related data) as published by the World Bank. An example of a redescription obtained by Mihelčić et al. (2017) is below:

$$[13.2 \leq POP_{14} \leq 15.2] \wedge [3.1 \leq MORT \leq 5.0] \wedge [0.0 \leq POP\_GROWTH \leq 0.5]$$

$$\sim [13.2 \leq E/I\_MiScManArt \leq 15.2] \wedge [28.0 \leq E\_MedSTechInMan \leq 40.0] .$$

This redescription describes seven countries (out of 199)—namely, Austria, Czech Republic, Germany, Italy, Poland, Slovenia, and Spain. On the one hand, these countries have a population consisting of between 13.2 % and 15.2 % of under 14 year olds, a mortality rate for under 5 year olds between 3.1 and 5.0 per 1000, and a slow population growth of 0.0 % to 0.5 %. On the other hand, these countries have an export–import ratio of miscellaneous manufactured articles between 13.2 and 15.2, and medium-skill technology-intensive manufactured goods make up between 28 % and 40 % of their export.

## 3.1.4 In Engineering

So far, our examples have used redescription mining as an exploratory data analysis method—which it of course is. But it can also be used for other purposes. Goel et al. (2010) used exact redescriptions over binary data to speed up sequential equivalence checking. Testing the equivalence of two logical circuits is a common problem in electrical engineering, where one typically first designs a circuit based on the logical requirements and then tries to optimize it, for example, by merging logical paths in order to reduce the number of transistors or surface area in the final circuit. Naturally, the optimized circuit must be equivalent to the original for the optimization to be valid.

A *sequential circuit* is a logical circuit that stores its output in flip-flops, from where it is fed as the input for the same circuit. An example of a sequential

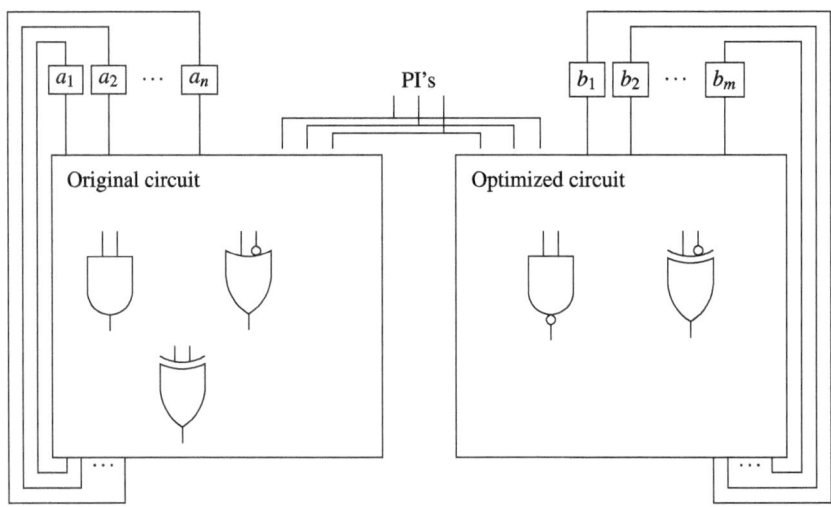

**Fig. 3.2** Example of a sequential circuit (left) and its optimized version (right). The primary inputs (PI's) provide the initial input, after which the output of every round is stored in flip-flops $a_1, a_2, \ldots, a_n$ (original circuit) or $b_1, b_2, \ldots, b_n$ (optimized circuit), from where it is fed as the input for the next round

circuit and its optimized version are presented in Fig. 3.2. Testing the equivalence of sequential circuits presents a hard problem: in a naive approach, each round of feeding the output as input would be modelled by copy-pasting the circuit after itself, making the circuit grow very large very quickly. On the other hand, the sequential structure of the circuit can create certain constraints on the possible inputs for the circuit, making certain input configurations impossible. A set of these configurations is called a *don't-care* set, and the equivalence checking can use the don't-care set to reduce the space of potential configurations it needs to validate.

Goel et al. (2010) use redescription mining to find candidate configurations to be included in the don't-care set. They run both the original and the optimized circuit for one round with a number of random inputs, storing the values of the flip-flops in an inputs-by-flip-flops binary table. They then use the BLOSOM algorithm (Zhao et al. 2006; see also Sect. 2.1.2) to mine the exact redescriptions from this matrix. These redescriptions provide candidates for illegal configurations; for instance, if BLOSOM finds a redescription $f_1 \wedge f_2 \equiv \neg f_3$, all configurations where flip-flop $f_3$ is true but either $f_1$ or $f_2$ is false are potentially illegal, as are those where all three flip-flops are true. To make sure that the candidate constraints are valid, Goel et al. (2010) validate them using a SAT solver. Valid constraints can be added to the model to make the final equivalence checking easier.

## 3.2 Relational Redescription Mining

Up to this point, our descriptions have characterized single objects, whether they were geographic locations, genes, medical patients, or political candidates. Next, we look at a redescription mining task where, instead of characterizing single objects in different ways, the goal is to find alternative descriptions of small groups of objects in terms of their individual properties and the relations that link them. In other words, we present in this section the relational variant of the redescription mining problem.

This problem variant could be useful in the exploration of knowledge bases and ontologies. It could help identify associations between the different relations that coexist in the data and possibly originate from different sources, going beyond the one-to-one mappings considered by most schema-matching approaches (Shvaiko and Euzenat 2005).

Relational redescription mining was introduced by Galbrun and Kimmig (2014). We begin with an example based on a data set from that work.

### 3.2.1 An Example of Relational Redescriptions

The data set used by Galbrun and Kimmig (2014) was extracted from the *Alyawarra Ethnographic Database*[5] and provides information about kinship terminology and family relationships within an Australian indigenous community. It can be seen as a pair of labelled graphs, sharing the same set of nodes. Nodes represent individuals from the community. One graph contains personal details and genealogic information. More precisely, the age and sex of the individuals are represented by node attributes, while parental and spousal relations between pairs of individuals are represented by edges, directed and undirected respectively, labelled with the type of the relation. The other graph contains information about the kinship terms individuals use for their relationships to other persons. This information is represented by directed edges, going from the speaker to the person referred to and labelled with the corresponding kinship term used. The entire data set contains almost 400 nodes and slightly over 24,000 edges in total for the two graphs.

Here, we only look at a small subset of that data set, which consists of 10 nodes and the relations between them from either of the two graphs. Figures 3.3 and 3.4 show the genealogic and kinship graphs that we will use in our example. The edges in the genealogic graph (see Fig. 3.3) are labelled with attribute gen which can take the value of either parent or spouse and is represented by an arrow and by a double line, respectively. The nodes are labelled with the attributes sex and age. The node attribute sex takes a value of either M, for male, or F, for female.

---

[5]http://www.culturalsciences.info/AlyaWeb/. Accessed 25 Oct 2017.

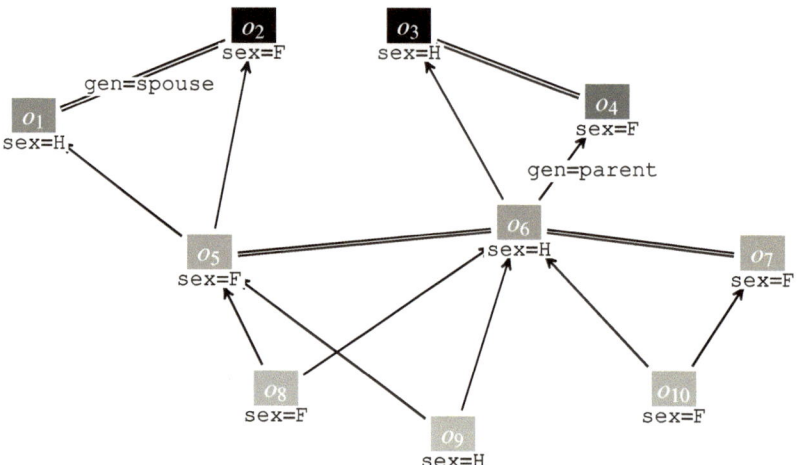

**Fig. 3.3** Genealogic graph from the *Alyawarra* data set as an example of a heterogeneous network used to mine relational redescriptions. Nodes represent individuals from the indigenous community, and edges represent parental and spousal relationships

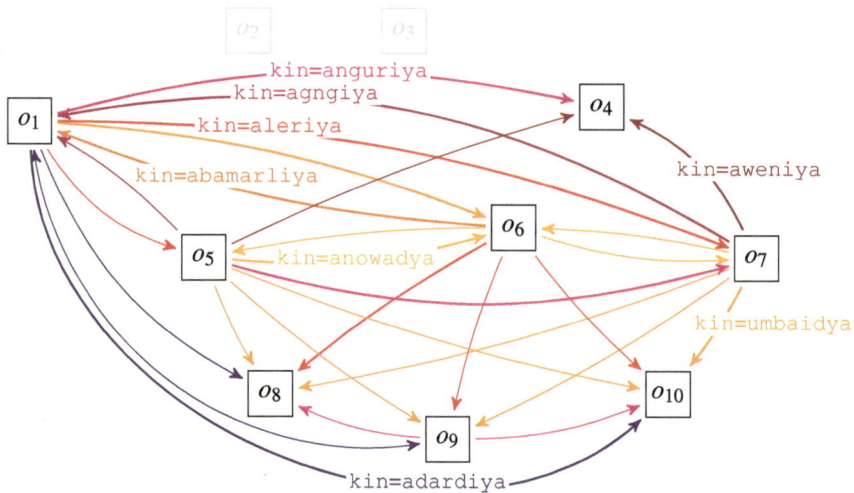

**Fig. 3.4** Kinship graph from the *Alyawarra* data set as an example of a heterogeneous network used to mine relational redescriptions. Nodes represent individuals from the indigenous community, and edges represent the kinship terms they use for one another

The node attribute `age` takes a numerical value indicated by the position of the node, with nodes at the bottom representing younger individuals. This attribute value is also indicated by the shade of the node, with darker nodes representing older individuals and black nodes $o_2$ and $o_3$ representing deceased individuals. The edges in the kinship graph (see Fig. 3.4) are labelled with the attribute `kin`,

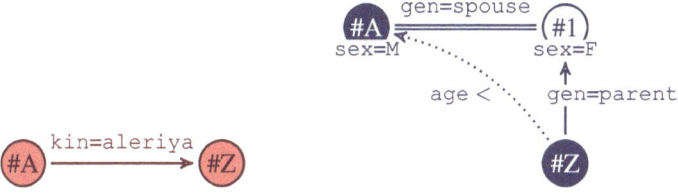

**Fig. 3.5** Genealogic and kinship graph queries over the *Alyawarra* data set

whose values represent the different kinship terms, each represented by arrows in a different colour. Kinship terms are not available for deceased individuals.

In this context, our goal is to find pairs of graph queries, such as the one shown in Fig. 3.5, that characterize roughly the same pairs of nodes in order to automatically discover definitions for the kinship terms. The nodes in graph queries are represented by circles to differentiate them from data nodes, represented by squares. The graph on the right-hand side of Fig. 3.5 represents a simple query that selects the pairs of individuals (#A, #Z) such that #A refers to #Z using the term *aleriya*. The graph on the left-hand side of Fig. 3.5 represents a more complex query that selects the pairs of individuals (#A, #Z) such that #A is male and older than #Z, and there exists a third individual, denoted as #1, who is female, the spouse of #A, and the parent of #Z. These two subgraphs form a relational redescription in the sense that the set of pairs of individuals that satisfy the first query largely matches the set of pairs of individuals that satisfy the second one. According to the glossary provided with the data, the term *aleriya* is used by male speakers to refer to a son or a daughter and by female speakers to refer to a son or a daughter of a brother. The pair of subgraphs shown in Fig. 3.5 encodes the first meaning of the term, that is, when used by male speakers. Hence, this redescription correctly recovers part of the definition of the term *aleriya*.

### 3.2.2 Formal Definition

Let us now formalize the definition of relational redescription mining. We use a formalism similar to the one introduced by Galbrun and Kimmig (2014) and Galbrun (2013). For a more in-depth discussion of the problem variant, the interested reader should refer to the original works.

Relational redescription mining takes relational data sets as input. In other words, the data sets considered contain relations between objects in addition to properties of individual objects, such as can be represented by hypergraphs. However, Galbrun and Kimmig (2014) restricted the problem to binary relations, considering only relations that involve two objects. Such a data set can be represented by a normal graph. The graph might contain both directed and undirected edges, and both nodes and edges are labelled with various attributes. As in the basic setup, the

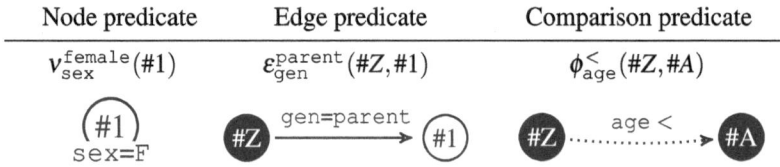

| Node predicate | Edge predicate | Comparison predicate |
|---|---|---|
| $v_{\texttt{sex}}^{\texttt{female}}(\#1)$ | $\varepsilon_{\texttt{gen}}^{\texttt{parent}}(\#Z, \#1)$ | $\phi_{\texttt{age}}^{<}(\#Z, \#A)$ |

**Fig. 3.6** Examples for the three types of predicates in relational redescription mining. For each type of predicate, an example from the *Alyawarra* data set is provided with the mathematical notation indicated on top and the equivalent graphical depiction underneath

set of attributes is divided into several views, and one can see the data set as consisting of a single graph with the attributes of all the different views or as consisting of several graphs, one for each view. In our example, the attributes are divided into two views. One view contains the personal information and genealogic relations, that is, it consists of node attributes sex and age as well as edge attribute gen. The other view contains the kinship information and consists only of attribute kin.

As in Sect. 1.2.2, we consider predicates over the attributes. Relational predicates are divided into three types: *node predicates*, *edge predicates*, and *comparison predicates*. In Fig. 3.6, we provide for each type of predicate an example from the *Alyawarra* data set, each denoted using the mathematical notation (top) and depicted graphically (bottom).

*Node predicates* test the value of an object attribute. Node predicates are, thus, the counterpart of the predicates defined in Sect. 1.2.2. The example node predicate in Fig. 3.6 (left) holds true for nodes for which the attribute sex takes the value female. In other words, this predicate selects as node #1 all nodes that represent female individuals.

*Edge predicates* test the value of an edge attribute. The example edge predicate in Fig. 3.6 (center) holds true for pairs of nodes that are linked by an edge for which the attribute gen takes the value parent. In other words, this predicate selects as nodes #Z and #1 all node pairs such that the individual represented by the second node is the parent of the individual represented by the first node.

Finally, *comparison predicates* compare the value of an attribute between two objects. The example comparison predicate in Fig. 3.6 (right) holds true for pairs of nodes such that the attribute age takes a smaller value for the first node than for the second node. Hence, this predicate selects as nodes #Z and #A all node pairs such that the individual represented by the first node is younger than the individual represented by the second node.

Note that for a given node or node pair relational predicates need to check that the attribute is defined before evaluating its value since it cannot be presumed, unlike in the table data model. For instance, the edge predicate $\varepsilon_{\texttt{gen}}^{\texttt{parent}}(\#Z, \#1)$ first needs to check that there exists an edge linking the two nodes and that it is labelled with the gen attribute before checking what value the attribute takes.

Graph queries are then obtained by taking a conjunction of predicates and marking some nodes as the query variables. In other words, a *graph query* is a definite clause of the form

$$q(\#A, \ldots, \#Z) \; = \; b_1 \wedge \ldots \wedge b_n \; ,$$

where the body elements $b_1, \ldots, b_n$ are node, edge, and comparison predicates, and the head variables $\#A, \ldots, \#Z$ are the *query variables*. The query variables represent the objects of interest in the query, the objects that it describes. For reasons of interpretability, we require the graph formed by edge predicates to connect the query variables. To differentiate between the query variables and the other nodes in the body of the query, we identify them, respectively, with letters and numbers. In graphics, we represent query variables with filled markers and use empty markers for other variables.

For instance, the genealogic graph query in Fig. 3.5 (right) consists of five predicates and can be denoted as

$$q(\#A, \#Z) \; = \; v_{\text{sex}}^{\text{male}}(\#A) \; \wedge \; \varepsilon_{\text{gen}}^{\text{spouse}}(\#A, \#1)$$

$$\wedge \; v_{\text{sex}}^{\text{female}}(\#1) \; \wedge \; \varepsilon_{\text{gen}}^{\text{parent}}(\#Z, \#1) \; \wedge \; \phi_{\text{age}}^{<}(\#A, \#Z) \; .$$

For a given query, one can look for matches between the nodes in the body of the query and the data nodes, such that each node in the body of the query is matched to a different data node while respecting the predicates. Such a match, where node $Y_j$ in the query maps to node $o_{i_j}$ in the data is called a *substitution* and is denoted as $\theta = \{Y_1/o_{i_1}, \ldots, Y_n/o_{i_n}\}$. It can be seen as a subgraph isomorphism when thinking in terms of graphs, and it is known as a $\theta_{OI}$-subsumption in inductive logic programming. An *answer substitution* is a substitution $\theta$ reduced to the query variables and the *support* of query $q$, denoted as supp($q$), is the set of all its distinct answer substitutions on the given data graph. Notice that in relational redescription mining, supp($q$) consist of object tuples and not just objects, as in standard redescription mining.

Going back to the above example, $\{\#A/o_3, \#1/o_4, \#Z/o_6\}$ is a substitution of $q(\#A, \#Z)$, and $(o_3, o_6)$ is the corresponding answer substitution. The support of this query in our example data set is

$$supp(q) = \{(o_1, o_5), (o_3, o_6), (o_6, o_8), (o_6, o_9), (o_6, o_{10})\} \; ,$$

The simple kinship graph query in Fig. 3.5 (left) can be denoted as

$$p(\#A, \#Z) \; = \; \varepsilon_{\text{kin}}^{\text{aleriya}}(\#A, \#Z) \; ,$$

and its support in our example data set is

$$supp(p) = \{(o_1, o_5), (o_1, o_7), (o_6, o_8), (o_6, o_9), (o_6, o_{10})\} \; .$$

As in the basic setup (1.6), we can use the Jaccard similarity together with some user-specified constant to define a similarity $\sim$ between graph queries. The Jaccard similarity between the support of the two graph queries $p$ and $q$ is

$$J(p, q) = \frac{4}{6} \; .$$

Heterogeneous networks and graph queries provide a new type of data $\mathcal{D}$ and a new query language $\mathcal{Q}$, resulting in the relational redescription mining problem variant when plugged into Definition 9.

**Definition 10 (Relational Redescription Mining)** Given data $\mathcal{D}$ consisting of heterogeneous networks, language $\mathcal{Q}$ of graph queries, similarity $\sim$, and other potential constraints, the goal of *relational redescription mining* is to find all valid redescriptions $(p_i, q_i)$ that also satisfy the other potential constraints.

In the relational redescription mining problem, as in the basic redescription mining problem (see Sect. 1.2.4), the other constraints might include restrictions on the support size, both from above and from below, as well as restrictions on the complexity of the graph query, limiting for instance the number of nodes, of edges, or of attributes that a query might involve. Also, to eliminate overly generic queries, it can be interesting to limit the number of distinct data nodes that a query node might map to. In this relational setting, evaluating the statistical significance of redescriptions is a rather complex issue and requires further investigation.

Galbrun and Kimmig (2014) presented an algorithm for mining relational redescriptions where the graph queries are limited to two *query variables*. They proposed a method to construct a graph query, given a list of target node pairs. It works by first finding simple labelled paths that appear frequently between the target pairs, that is, short paths that link as many node pairs as possible among the target pairs. Then, those simple paths are filtered and combined into more complex graph queries, based on their occurrences in the data set. This relational query miner forms the basis of an alternating scheme that uses the support of the query found at one step as the target for building a new query in the following step.

## 3.3  Storytelling

Storytelling is an extension of redescription mining that was initially proposed by Ramakrishnan et al. (2004). Current storytelling algorithms all follow the same framework based on the A* heuristic. Nonetheless, this framework has proven to be a versatile approach, applicable to different domains.

### 3.3.1 Definition and Algorithms

The goal of storytelling is to build a *story* between two queries.

**Definition 11 (Stories and Storytelling)** Given data $\mathcal{D}$, query language $\mathcal{Q}$, similarity $\sim$, and two queries $q_s$ and $q_t$ from $\mathcal{Q}$, a *story* between $q_s$ and $q_t$ is a sequence $q_s = q_1, q_2, q_3, \ldots, q_k, q_{k+1} = q_t$, of queries from $\mathcal{Q}$ such that for all $i = 1, 2, \ldots, k$ the pair $(q_i, q_{i+1})$ is a valid redescription, except that their views do not have to be disjoint (i.e. $q_i \sim q_{i+1}$, but it is possible that views$(q_i) \cap$ views$(q_{i+1}) \neq \emptyset$). The *length* of the story is $k$. The goal of *storytelling* is to find the shortest story between given queries $q_s$ and $q_t$.

The following example is by Kumar et al. (2008).

*Example 6* Consider the game of morphing words by changing a few letters from one word to obtain another word. The goal is to reach the target word through a succession of morphs between valid words. This game can be modelled as a storytelling task. The attributes are English words—in this case, all words with five letters—and the entities are *(letter, position)* pairs. In our query language, we only allow singleton queries. For example, the query $q = $ 'booth' would have supp$(q) = \{(b, 1), (o, 2), (o, 3), (t, 4), (h, 5)\}$. The amount of letters that each morph is allowed to change is controlled by the threshold on the Jaccard similarity. For example, looking for a story from $q_s = $ 'booth' to $q_t = $ 'flash' and setting the threshold so that two letters can be changed, Kumar et al. (2008) obtained the following:

booth $\sim$ boats $\sim$ beams $\sim$ deads $\sim$ grads $\sim$ grade $\sim$ craze $\sim$ crash $\sim$ flash.

The requirement for disjoint views is dropped for the redescriptions in a story. Indeed, there is no risk of tautological queries as the subsequent redescriptions aim to be more similar with $q_t$. Besides, requiring that the subsequent views are disjoint would mean that the second to last query cannot share attributes with $q_t$.

Kumar et al. (2008) propose a framework for storytelling. The framework is based on the A* heuristic and redescription mining and is presented in Algorithm 3.1. The framework operates in steps. At every step, it considers a query $q$. If this query is sufficiently similar to the target, we have found the story and return it (line 5). Otherwise, $b$ new redescriptions $q \sim q'$ are mined (lines 7–11). For each of the candidate queries $q'$, we calculate how many steps, at minimum, it will take to reach query $q_t$ (line 9). This is used to sort the candidates in the priority queue: the queries $q'$ are entered to the priority queue with a key that is the sum of the current number of steps taken and the minimum number of steps still needed. Together with the query, we also store the number of steps taken so far and the current partial story (line 10).

---

**Algorithm 3.1** Framework for storytelling

---

**Input:** Data $\mathcal{D}$, query language $\mathcal{Q}$, similarity $\sim$, starting query $q_s \in \mathcal{Q}$, target query $q_t \in \mathcal{Q}$, and
    branching parameter $b$.
**Output:** A story $(q_s, q_2, q_3, \ldots, q_k, q_t)$ or $\emptyset$.
 1: $Q \leftarrow$ PriorityQueue();   $Q$.put$(0, q_s, 0, \emptyset)$
 2: **while** $Q$ is not empty **do**
 3:    $(e, q, k, h) \leftarrow Q$.pop()
 4:    **if** $q \sim q_t$ **then**
 5:       **return** $h \sim q \sim q_t$
 6:    **end if**
 7:    **for** $i = 1, \ldots, b$ **do**
 8:       find a query $q' \in \mathcal{Q}$ such that $q \sim q'$
 9:       $m \leftarrow$ the minimum number of steps from $q'$ to $q_t$
10:       $Q$.put$(k + m, q', k + 1, h \sim q')$
11:    **end for**
12: **end while**
13: **return** $\emptyset$

---

If the priority queue becomes empty without finding any story, the framework
returns an empty set (line 13). This can happen even if a story between $q_s$ and $q_t$
exists: if the user-supplied parameter $b$ is too small, the A* heuristic cannot search
the neighbourhoods of the queries properly and can miss a valid story.

Kumar et al. (2008) use the CARTwheels algorithm (Ramakrishnan et al. 2004;
see also Sect. 2.2) for finding the queries and they only consider binary data. In
principle, any other query-finding algorithm could be used, not restricted to binary
data. This would, however, require one to find a way to estimate the minimum
number of steps that are still needed, as this estimator is the key to the A* heuristic
and very dependent on the chosen algorithm (and query language).

Another method, as used by Hossain et al. (2012b) (see also Sect. 3.3.2) is to use
as the proxy for quality not the number of steps but rather the distance between the
current query and the target query. For this, the distance function has to admit the
triangle inequality.

*Example 7* To see how the algorithm works, consider the following example,
adapted from Kumar et al. (2008). The data are stored in a single Boolean data
table

$$
\mathbf{D} = \begin{array}{c} \\ e_1 \\ e_2 \\ e_3 \\ e_4 \\ e_5 \\ e_6 \end{array} \begin{array}{c} \begin{array}{cccccc} \alpha & \beta & \gamma & \delta & \varepsilon & \zeta \end{array} \\ \left( \begin{array}{cccccc} 1 & 1 & 0 & 0 & 0 & 0 \\ 0 & 1 & 1 & 0 & 0 & 0 \\ 0 & 1 & 0 & 1 & 0 & 0 \\ 0 & 0 & 1 & 0 & 0 & 0 \\ 0 & 0 & 0 & 1 & 1 & 0 \\ 0 & 0 & 0 & 0 & 0 & 1 \end{array} \right) \end{array},
$$

with the source and target queries being $q_s = \alpha$ and $q_t = \varepsilon$, respectively. We
define the similarity threshold so that $p \sim q$ if $J(p, q) \geq 1/2$. The source query

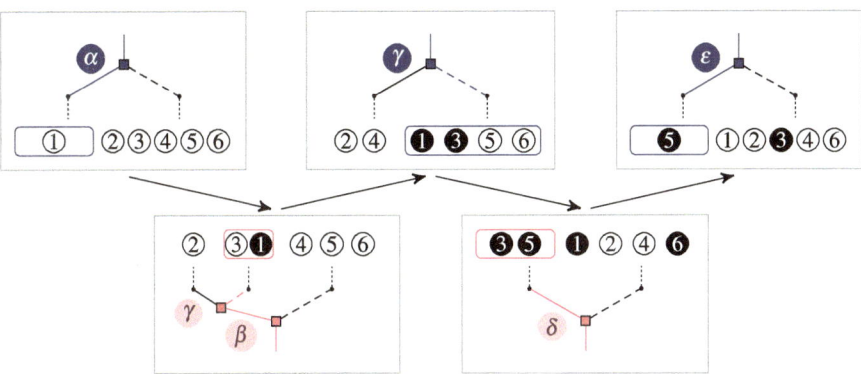

**Fig. 3.7** Tree diagrams depicting the queries from Example 7. Odd-numbered and even-numbered queries are shown in the top and bottom half, respectively. In each diagram, we draw the entities as filled or empty circles depending on whether they carry a positive or a negative label, that is, depending on whether or not they belong to the support of the previous query, respectively. Entities in the support of the current query are enclosed in coloured bins

has support $\text{supp}(q_s) = \{e_1\}$, and the target query has support $\text{supp}(q_t) = \{e_5\}$, and so the first induced query needs to have $e_1$ and one other entity in its support. Let that query be $q_2 = \beta \wedge \neg \gamma$ with support $\{e_1, e_3\}$. This query does not yet have any overlap in support with $q_t$, but the next query $q_3 = \neg \gamma$ does have $e_5$ in its support $\text{supp}(q_3) = \{e_1, e_3, e_5, e_6\}$. Unfortunately, $q_3 \not\sim q_t$, as $J(q_3, q_t) = 1/4$. So the algorithm continues, inducing query $q_4 = \delta$ with support $\{e_3, e_5\}$. This query has a Jaccard similarity of $1/2$ to $q_t$, completing our storytelling chain of queries as

$$q_s = \alpha \sim \beta \wedge \neg \gamma \sim \neg \gamma \sim \delta \sim \varepsilon = q_t \,.$$

The queries are visualized using the tree diagrams in Fig. 3.7. For the colours in the visualizations, the odd-numbered queries are always considered as the left-hand queries and the even-numbered queries as the right-hand queries. They are shown in the top and bottom half of the figure, respectively. In each diagram, we draw the entities as filled or empty circles, depending on whether they carry a positive or a negative label, that is, depending on whether or not they belong to the support of the previous query. Entities in the support of the current query are enclosed in coloured bins.

### 3.3.2 Applications

Hossain et al. (2012b) use storytelling to connect PubMed articles, and more generally, to connect molecules using stories between PubMed articles that mention the molecules in their abstracts. In their framework, articles in the PubMed publication repository[6] are the entities, and the terms appearing in their abstracts are the attributes. The attribute values indicate the weight of the term in the document and are calculated based on a variant of the standard tf–idf (term frequency–inverse document frequency) score.

The first step in the analysis pipeline of Hossain et al. (2012b) is that the user has to choose a set of input molecules and a set of output molecules based on her interests. Next, PubMed documents that mention these molecules are crawled, and the $(q_s, q_t)$ query pairs are formed so that each query selects exactly one article that mentions exactly one of the input or output molecules. In addition, the abstract of the article selected by $q_s$ should have no common terms with the abstract of the article selected by $q_t$.

The storytelling algorithm (Algorithm 3.1) is now run with all of the $(q_s, q_t)$ pairs. The query language is restricted so that each query can select only one document. The distance between two documents $a$ and $b$ is calculated using the Soergel distance,

$$d(a,b) = \frac{\sum_t |w_{t,a} - w_{t,b}|}{\sum_t \max\{w_{t,a}, w_{t,b}\}} \; ,$$

where the sums run over all terms $t$ in the data and $w_{t,i}$ is the weight for term $t$ in document $i$. Two documents are considered similar (enough) if their Soergel distance is below the user-supplied distance threshold $\tau$.

In addition to the distance threshold $\tau$, Hossain et al. (2012b) add another constraint, which they call the *clique size*. If we consider a graph where the nodes are the documents, and there is an edge between nodes $a$ and $b$ if $d(a,b) \leq \tau$ for the distance threshold $\tau$, setting the clique size to $k$ means that any selected document must be part of a clique of size at least $k$ in this graph. Hence, any two consecutive documents in the story are in the same clique.

To find the queries, Hossain et al. (2012b) first consider the binary matrix $([w_{t,i} > 0])_{t,i}$, that is, the matrix that indicates whether term $t$ is present in document $i$. They run the CHARM-L algorithm (Algorithm 2.2) on this matrix in order to find the concept lattice of documents. From this, they select the actual documents (i.e. queries) to be added to the priority queue. The goal of Hossain et al. (2012b) is not to find the shortest stories, but to find some stories, and the A* heuristic is used to find documents that are close to the target document. The lower-bound on the distance is calculated as the Soergel distance between the current document and the target document, since the Soergel distance admits the triangle inequality.

Running the storytelling algorithm for each $(q_s, q_t)$ pair results in many stories. Hossain et al. (2012b) filter this initial set of stories based on the $p$-value of the stories, their coherency, and the context overlap between the articles. An example of a story found by Hossain et al. (2012b) is depicted in Table 3.1.[7]

---

[6]https://www.ncbi.nlm.nih.gov/pubmed/. Accessed 25 Oct 2017.

[7]Story adapted from https://bioinformatics.cs.vt.edu/connectingthedots/stories.html, case study 3 (accessed 25 Oct 2017).

**Table 3.1** Example story from *pyruvate kinase* to *glutamine* (Hossain et al. 2012b)

| PubMed ID | Molecule | Article title |
| --- | --- | --- |
| 16511150 | pyruvate kinase | Crystallization and preliminary X-ray analysis of pyruvate kinase from Bacillus stearothermophilus |
| 3688482 | adenosine diphosphate | Metabolism of round spermatids: kinetic properties of pyruvate kinase |
| 9441794 | adenosine triphosphate | Phosphoenolpyruvate prevents the decline in hepatic ATP and energy charge after ischemia and reperfusion injury in rats |
| 16552804 | glutamine | Alanyl-glutamine dipeptide inhibits hepatic ischemia-reperfusion injury in rats |

Hossain et al. (2012a) use an approach similar to the above one to analyse entity networks for intelligence purposes. Later, Wu et al. (2014) presented another approach based on storytelling for that task. In this data intelligence task, they consider case studies developed at the Joint Military Intelligence College (USA). Each case study contains a collection of fictional intelligence reports. Binary relations are extracted from these documents and turned into multi-relational binary data. This can be modelled as a chain of binary matrices $B_1, B_2, \ldots$, so that two consecutive matrices, $B_i$ and $B_{i+1}$, correspond to the same attributes or the same entities in one dimension. For example, in Fig. 3.8, we can see how the phone numbers-by-dates matrix can be connected to the phone numbers-by-addresses matrix, which can be connected to the persons-by-addresses matrix, which can be connected to the persons-by-corporations matrix, and so on.

Notice how this approach allows us to transpose the data for making the connection and the story. If $q_1$ is a query over the phone numbers-by-dates matrix, with dates as attributes, then $q_2$ could be a query over the phone numbers-by-addresses matrix, with addresses as attributes. For both queries, the support would consist of a set of phone numbers. But for the next query, $q_3$, over the persons-by-addresses matrix, the attributes would be the persons, and the addresses would be the entities. To compare the supports of queries $q_2$ and $q_3$, we transpose query $q_2$, turning its support into a new query. The support of the transposed query will be $\mathrm{dscr}(\mathrm{supp}(q_2))$, a set of addresses (see Sect. 2.1). Such transposition is possible as we only consider monotone conjunctive queries over binary data.

The goal of Wu et al. (2014) is to 'uncover a plot', that is, to find a surprising connection between the entries in such binary matrices. The connections, or plots, are modelled as chains of biclusters, that is, stories where the query language comprises monotone conjunctive queries over binary attributes. The surprisingness of the stories is measured with respect to a constrained maximum entropy distribution, where the constraints, as in Sect. 1.2.6, encode what is already known about the data. Wu et al. (2014) use a simple greedy heuristic for building the stories, similar to the A* heuristic of Algorithm 3.1.

An example plot uncovered by Wu et al. (2014) is shown in Fig. 3.9. The story goes from the locations-by-persons matrix to the months-by-locations matrix with

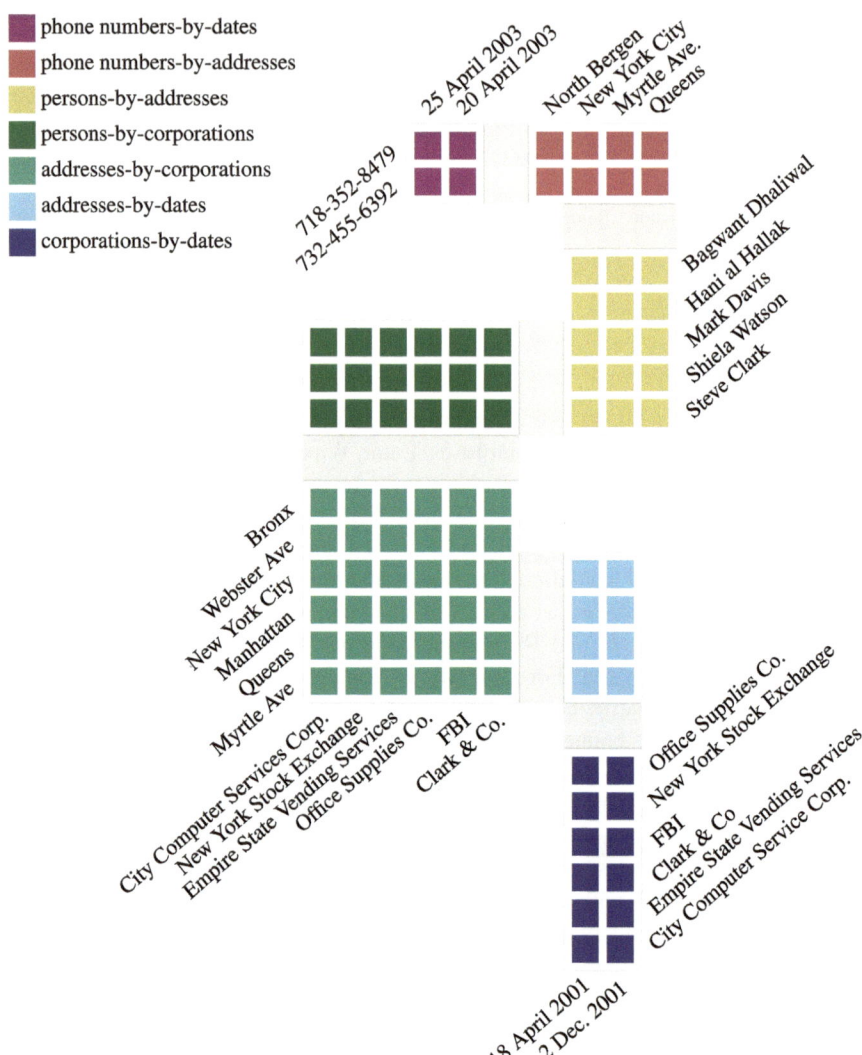

**Fig. 3.8** Example data for uncovering the plot. Figure adapted from Wu et al. (2014)

successive biclusters shown in different colours. Wu et al. (2014) present the actual plot as follows:

> **Fahd al Badawi, Boris Bugarov, Adnan Hijazi, Jose Escalante,** and **Saeed Hasham** coordinate with each other to recruit **Al Qaeda** field agents to transport biological agents to USA via **Holland Orange Shipping Lines.**

The entities in the uncovered plot that are part of the true plot are set in boldface font in Fig. 3.9,

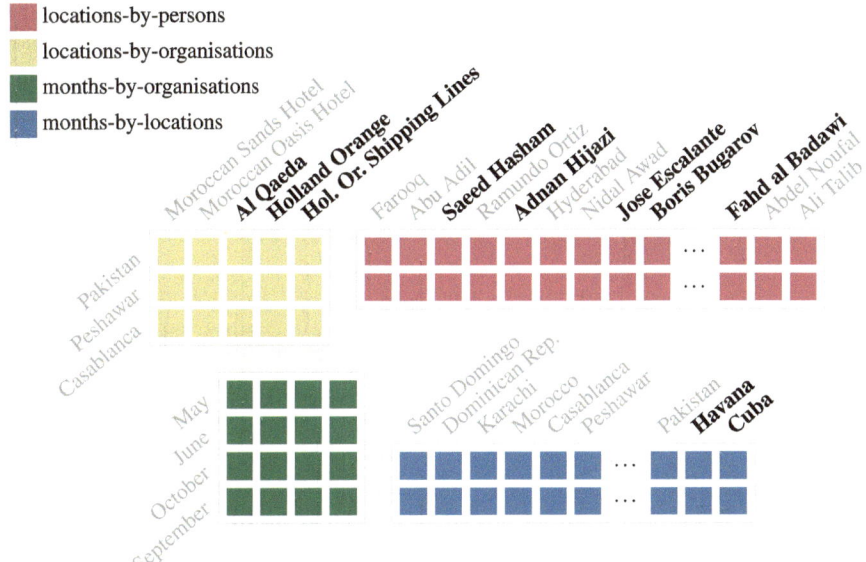

**Fig. 3.9** Example uncovered plot. Entities in boldface font are part of the actual plot (see text). Figure adapted from Wu et al. (2014)

As can be seen from Fig. 3.9, the biclusters contain many more entities than the plot, and the Holland Orange Shipping Lines appear twice under two different names. Having false positives (entities unrelated to the actual plot) is probably preferred over false negatives in this application domain, provided that there are not too many of them.

## 3.4 Future Work: Richer Query Languages

In this closing section, we discuss directions for future work, focusing on the development of richer query languages. By allowing us to take into account time or graph structures, for instance, richer query languages facilitate the application of redescription mining to data sets—such as in the fields of finance or chemistry—where the additional properties are important.

### 3.4.1 Time-Series Redescriptions

In some cases, the temporal dimension of the data is important. In finance, for instance, if one follows the value of selected stocks over the course of some days or months, the data will take the form of a time series, that is, a collection of data points indexed in time. Similarly, in order to analyse and optimize production processes in an industrial setting, the reading of various sensors on the machinery might be recorded regularly, resulting in time-series data.

Depending on the application and the goal of the analysis, one might consider each time point as constituting one entity, while each stock or sensor is represented by one attribute; vice versa, one might take the time points as the attributes, while stocks and sensors constitute the set of entities.

When each time point constitutes one entity, we can use conditional redescriptions to create redescriptions that only hold (accurately) within some (contiguous) span of time. If $t$ is a (pseudo-)attribute that encodes the time, conditional redescriptions such as $p \sim q \mid [a \leq t \leq b]$ would restrict the redescription to being evaluated only between time $a$ and time $b$. To require that the redescriptions hold *only* within that time span, we could further enforce that the redescription must have a very low similarity when evaluated in the complement of the condition. This would allow us to find *temporally autocorrelated* redescriptions, that is, if they start holding true at time point $a$, they (mostly) hold true until time point $b$, after which they do not hold true.

Another approach, to find recurrent events, would be to have attributes corresponding to the time in one view and the original attributes in another view. The time attributes could encode information such as the hour, the day of the week, the day and the month, or the year of the time point. This way, the redescriptions would characterize the entities using the original attributes and the time when the entities happened, for example,

$$[\text{alcohol sales} \geq \$10{,}000] \sim [\text{day} \in \{\text{Fri, Sat}\}] \vee [\text{date} \in \{31 \text{ April}, 31 \text{ Dec}\}] \,,$$

indicating that in a (hypothetical) shop, the alcohol sales peak on Fridays and Saturdays and in the eves of May Day and New Year, irrespective of which day of the week these holidays fall on.

When looking at time series where the attributes represent time points, one might be less interested in the actual value at some given point in time than in the variation of the value over some time span. Of course, one could compute the differences in value between chosen time points, add them as new attributes, and mine this extended data.

A first drawback of this approach, however, is that it can lead to an explosion in the number of attributes. To avoid adding attributes needlessly, it might be interesting to let the algorithm generate attributes on-the-fly when it needs them, rather than have to extend the data as a preprocessing step.

This might be useful not only for time series but also for other types of data, more generally. For instance, to find bioclimatic niches, it might be interesting to consider trait distributions rather than species presence (see Sect. 3.1.2). In addition to a collection of raw variables, the user could specify as the input to the algorithm some operators (e.g. difference, average, minimum, etc.) that, when applied on subsets of the raw variables, allow us to obtain meaningful composite variables. Of course, such an automated feature engineering approach would result in a much expanded search space. Hence, efficient exploration and pruning strategies need to be devised in order for such an approach to be practical.

A second drawback of adding composite attributes is that it does not allow us to take into account the dependencies between the created attributes in a straightforward manner. For instance, if one extends a time-series data set by generating a collection of attributes $\{v_{ij}\}$, where $v_{ij}$ represents the variation in value between time points $t_i$ and $t_j$, it is hardly of any interest to find a redescription of the form

$$[a \leq v_{ij} \leq a'] \wedge [b \leq v_{jk} \leq b'] \sim [a + b \leq v_{ik} \leq a' + b'] .$$

To avoid generating such tautological and uninteresting redescriptions, the dependencies between the attributes need to be encoded in the data and taken into account by the algorithm during the mining process.

Again, this could be useful beyond time-series data. If there exist known dependencies between some attributes or if the categories of some attribute are organized in a hierarchy, for instance, it would be beneficial to take such information into account, to improve both the efficiency of the mining process as well as the quality of the obtained redescriptions.

## 3.4.2 Subgraph Redescriptions

In sequences, where the data consist of successive occurrences of various events, event occurrences might be associated to individual timestamps or be ordered only with respect to one another. In the first case, the temporal dimension is of importance, but in the latter, it is secondary to the order relationship, if not entirely absent. For instance, one might consider the different activities undertaken by an individual during the course of a few weeks or a month as a sequence of events, each one associated to the time when the corresponding activity started. Snippets of text and DNA sequences are other examples of sequential data, ones without a temporal dimension.

Similarly to the model with a single data table (see Definition 7), one could consider a data set where each entity is associated to a sequence. The goal of redescription mining, in this case, would be to identify pairs of subsequences such that the two subsequences occur in almost the same set of entities. For instance, one could consider individuals, each one represented by one entity, associated to the corresponding sequence of activities. A redescription would then identify a pair of activity routines that provide two different characterizations of individuals sharing a similar lifestyle.

At yet a higher level of complexity, we could consider graph data sets, where each entity is associated with one or several small graphs instead of with a sequence.

For instance, in chemistry, one might consider the structure of molecules, as represented in the form of graphs. One might then be interested in looking for associations between the presence of some substructures and some properties of the molecule, such as being toxic or mutagenic, reacting to a particular protein, etc. This problem could be formulated as a redescription mining task, with a graph data set on the one hand and a data table on the other hand. More specifically, each entity would stand for a molecule, associated with a graph representing its structure, on the one hand, and with values for various attributes collected as a row in a table, on the other hand. This yields a setting similar to the model with two data tables (see Definition 7), one of which is substituted with a collection of graphs. One description would require the occurrence of a particular subgraph, selecting as its support those molecules that contain the specific fragment, while the other description would be a query over the attributes, selecting as its support those molecules that have the specified properties.

In order to extend redescription mining to handle sequences and graphs, the procedure for building the descriptions, or in other words, for mining sub-sequences or sub-graphs, can draw inspiration from existing techniques such as gSpan by Yan and Han (2002) or the Gaston tool by Nijssen and Kok (2005). In particular, existing algorithms could be used to mine patterns, which would then be paired based on support, following a mine-and-pair approach (see Sect. 2.1). However, this is probably not feasible in practice for any but the smallest data sets, due to the large amount of patterns that might be generated. A careful adaptation will be needed to build efficient algorithms for mining redescriptions involving complex structures such as sequences and graphs, when the non-tabular data set is considered alone as well as combined with a tabular data set.

### 3.4.3  Multi-Query and Multimodal Redescriptions

Similar to storytelling (Sect. 3.3), we can also consider extending redescription mining to more than two queries, that is, to *multi-query* redescription mining.[8] Unlike storytelling, multi-query redescriptions have no user-defined source or target queries, but the number of queries should be pre-determined. For example, if the attributes are divided into three views $V_1$, $V_2$, and $V_3$, we can try to find triples of queries $(q_1, q_2, q_3)$ such that views$(q_i) = \{V_i\}$ for all $i$. Naturally, this approach extends to arbitrary many queries and views.

Being able to connect more than two views could be beneficial in some applications. Consider the ecological niche application in Chap. 1, for example. Instead of species and bioclimatic variables as the two views, we could have carnivores, herbivores, and bioclimatic variables as three views. This would allow us to find connections, not just between species and climate, but also between carnivorous and herbivorous species, in relation to climate.

---

[8]The result should perhaps be called 'tridescription' or 'multi-description', though.

Perhaps the biggest problem with multi-query redescription mining is the definition of the distance between the queries. The problem here is not that there does not exist meaningful distance measures; rather, the problem is that there exist many meaningful ways to extend the measures, but none of them are clearly better than the others. To illustrate this, let us consider the case of three queries and different ways to define the three-way distance measure $d(q_1, q_2, q_3)$.

To begin, we can consider the pairwise distances

$$d_{1,2} = d(q_1, q_2), \quad d_{1,3} = d(q_1, q_3), \quad \text{and} \quad d_{2,3} = d(q_2, q_3) .$$

We can base the three-way distance measure on any function over these pairwise distances. For example, if we want to have all of the pairwise distances small, we can define

$$d(q_1, q_2, q_3) = \max\{d_{1,2}, d_{1,3}, d_{2,3}\} . \tag{3.1}$$

Alternatively, if our pairwise distance admits the triangle inequality, we can just bound the total distance and define

$$d(q_1, q_2, q_3) = d_{1,2} + d_{2,3} , \tag{3.2}$$

since $d_{1,3} \leq d_{1,2} + d_{2,3}$. Naturally, choosing any other pair of the distances would work, as well.

Especially if the distance is not a metric, we can consider the average pairwise distance

$$d(q_1, q_2, q_3) = \frac{d_{1,2} + d_{1,3} + d_{2,3}}{3} , \tag{3.3}$$

which would allow one larger distance, as long as the other two are small enough. For metric distances this might not be that useful a definition, as the triangle inequality prevents large variations in the distances. If there are more than three queries, though, the average could be useful even for metric distances.

Instead of working with the pairwise distances, we can define the three-way distance directly using the supports. For example, we can define the three-way Jaccard similarity coefficient as

$$J(q_1, q_2, q_3) = \frac{|\text{supp}(q_1) \cap \text{supp}(q_2) \cap \text{supp}(q_3)|}{|\text{supp}(q_1) \cup \text{supp}(q_2) \cup \text{supp}(q_3)|} . \tag{3.4}$$

None of these distance measures are inherently better than the other, and the choice between them must, to an extent, be based on the use case. In addition, one must also consider whether it is possible to design an algorithm that finds multi-query redescriptions under the chosen distance measure.

Instead of multiple independent views, we can also consider *multimodal* data and redescriptions. Instead of having a binary relation between entities and attributes, we can have ternary or multi-ary relations. For example, continuing with the bioclimatic niche finding example, the data could contain the species' habitats and bioclimatic variables at different points of time. The redescriptions would then link certain species to certain bioclimatic variables at certain points of time.

To formalize this concept, consider the table-based data model. In the standard niche finding example, we have two tables (or matrices), $\mathbf{D}_1$ (locations-by-species) and $\mathbf{D}_2$ (locations-by-bioclimatic variables). If we add the time dimension, we instead have two *tensors*, $\mathcal{T}_1$ (locations-by-species-by-time) and $\mathcal{T}_2$ (locations-by-bioclimatic variables-by-time). The predicates for these queries could fix two of the modes and the support would be a set of entities where the queries hold. An example redescription could be

$$\text{polar bear}@[\text{year} \leq 1970] \sim [-7.07 \leq t_5 \leq -3.38]@[\text{year} \leq 1970] \, ,$$

stating that the connection between the polar bear and temperature must hold for all years prior to, and including 1970.

Including time as the third mode highlights the connection to time-series redescription mining (see Sect. 3.4.1). However, the third mode does not have to be time. Knowledge bases, such as YAGO (Suchanek et al. 2007), store vast amounts of knowledge in the *Resource Description Framework* (RDF)[9] format. An RDF data set consists of subject–predicate–object (or $(s, p, o)$) triples, and it is often treated as a directed labelled graph, where subjects and objects are the vertices, and predicates are encoded as directed labelled edges between them, similarly to the data in relational redescription mining (see Sect. 3.2). Instead of as a graph, we can also treat an RDF data set as a three-way binary tensor (Metzler and Miettinen 2015a,b), with one mode for subjects, one for objects, and one for predicates. This tensor can be used as a data set for multimodal redescription mining; the redescriptions could, for example, be used to find (almost) synonymous relations. A redescription

$$\text{studiedIn} \sim \text{graduatedFrom}$$

would indicate that (almost) all pairs $(x, y)$ such that $x$ studied in $y$ are also such that $x$ graduated from $y$ (and vice versa).

This example also illustrates one complication of multimodal redescriptions: what are the entities and what are the attributes? In the first example, the entities were the locations, and the attributes were the species–time and bioclimatic variable–time pairs. In the current example, though, the attributes are the relations (or predicates), and the entities are the subject–object pairs. We might also want to allow for more complex queries, such as

$$[?x \text{ graduatedFrom } *] \sim [?x \text{ hasDegree } *] \, ,$$

---

[9]http://www.w3.org/TR/rdf-syntax-grammar. Accessed 25 Oct 2017.

which would mean that those who graduated (from anywhere) hold some degree. Indeed, for queries over RDF data, we might want to use (a subset of) the SPARQL query language,[10] but for data with more than three modes, even more complex query languages might be needed.

# References

Gaidar D (2015) Mining redescriptors in Staphylococcus aureus data. Master's thesis, Universität des Saarlandes, Saarbrücken

Galbrun E (2013) Methods for redescription mining. PhD thesis, Department of Computer Science, University of Helsinki

Galbrun E, Kimmig A (2014) Finding relational redescriptions. Mach Learn 96(3):225–248, https://doi.org/10.1007/s10994-013-5402-3

Galbrun E, Miettinen P (2012) From black and white to full color: Extending redescription mining outside the Boolean world. Stat Anal Data Min 5(4):284–303, https://doi.org/10.1002/sam.11145

Galbrun E, Miettinen P (2016) Analysing political opinions using redescription mining. In: IEEE International Conference on Data Mining Workshops, pp 422–427, https://doi.org/10.1109/ICDMW.2016.0066

Galbrun E, Tang H, Fortelius M, Žliobaitė I (2017) Computational biomes: The ecometrics of large mammal teeth. Palaeontol Electron. Submitted

Goel N, Hsiao MS, Ramakrishnan N, Zaki MJ (2010) Mining complex Boolean expressions for sequential equivalence checking. In: Proceedings of the 19th IEEE Asian Test Symposium (ATS'10), pp 442–447, https://doi.org/10.1109/ATS.2010.81

Hossain MS, Butler P, Boedihardjo AP, Ramakrishnan N (2012a) Storytelling in entity networks to support intelligence analysts. In: Proceedings of the 18th ACM SIGKDD International Conference on Knowledge Discovery and Data Mining (KDD'12), pp 1375–1383, https://doi.org/10.1145/2339530.2339742

Hossain MS, Gresock J, Edmonds Y, Helm RF, Potts M, Ramakrishnan N (2012b) Connecting the dots between PubMed abstracts. PLoS ONE 7(1):1–23, https://doi.org/10.1371/journal.pone.0029509

Kohonen T (1989) Self-organization and associative memory. Springer, New York

Kumar D (2007) Redescription mining: Algorithms and applications in bioinformatics. PhD thesis, Department of Computer Science, Virginia Polytechnic Institute and State University

Kumar D, Ramakrishnan N, Helm RF, Potts M (2008) Algorithms for storytelling. IEEE Trans Knowl Data En 20(6):736–751, https://doi.org/10.1109/TKDE.2008.32

van Leeuwen M, Galbrun E (2015) Association discovery in two-view data. IEEE Trans Knowl Data Eng 27(12):3190–3202, https://doi.org/10.1109/TKDE.2015.2453159

Metzler S, Miettinen P (2015a) Join size estimation on Boolean tensors of RDF data. In: Proceedings of the 24th International Conference on the World Wide Web (WWW'15), pp 77–78, https://doi.org/10.1145/2740908.2742738

Metzler S, Miettinen P (2015b) On defining SPARQL with Boolean tensor algebra. https://doi.org/10.1145/2740908.2742738, arXiv:1503.00301

Mihelčić M, Džeroski S, Lavrač N, Šmuc T (2017) A framework for redescription set construction. Expert Syst Appl 68:196–215, https://doi.org/10.1016/j.eswa.2016.10.012

Mihelčić M, Šimić G, Babić-Leko M, Lavrač N, Džeroski S, Šmuc T (2017) Using redescription mining to relate clinical and biological characteristics of cognitively impaired and Alzheimer's disease patients. arXiv:1702.06831

---

[10]http://www.w3.org/TR/rdf-sparql-query. Accessed 25 Oct 2017.

Mihelčić M, Džeroski S, Lavrač N, Šmuc T (2016) Redescription mining with multi-target predictive clustering trees. In: Proceedings of the 4th International Workshop on the New Frontiers in Mining Complex Patterns (NFMCP'15), pp 125–143, https://doi.org/10.1007/978-3-319-39315-5_9

Nijssen S, Kok JN (2005) The Gaston tool for frequent subgraph mining. Proceedings of the International Workshop on Graph-Based Tools (GraBaTs 2004) 127(1):77–87, https://doi.org/10.1016/j.entcs.2004.12.039

Pearson RG, Dawson TP (2003) Predicting the impacts of climate change on the distribution of species: Are bioclimate envelope models useful? Glob Ecol Biogeogr 12:361–371, https://doi.org/10.1046/j.1466-822X.2003.00042.x

Phillips SJ, Anderson RP, Schapire RE (2006) Maximum entropy modeling of species geographic distributions. Ecol model 190(3):231–259, https://doi.org/10.1016/j.ecolmodel.2005.03.026

Ramakrishnan N, Zaki MJ (2009) Redescription mining and applications in bioinformatics. In: Chen J, Lonardi S (eds) Biological Data Mining, Chapman and Hall/CRC, Boca Raton, FL

Ramakrishnan N, Kumar D, Mishra B, Potts M, Helm RF (2004) Turning CARTwheels: An alternating algorithm for mining redescriptions. In: Proceedings of the 10th ACM SIGKDD International Conference on Knowledge Discovery and Data Mining (KDD'04), pp 266–275, https://doi.org/10.1145/1014052.1014083

Roweis ST, Saul LK (2000) Nonlinear dimensionality reduction by locally linear embedding. Science 290(5500):2323–2326, https://doi.org/10.1126/science.290.5500.2323

Shvaiko P, Euzenat J (2005) A survey of schema-based matching approaches. J Data Semantics IV 3730:146–171, https://doi.org/10.1007/11603412_5

Singh J, Kumar D, Ramakrishnan N, Singhal V, Jervis J, Garst JF, Slaughter SM, DeSantis AM, Potts M, Helm RF (2005) Transcriptional response of Saccharomyces cerevisiae to desiccation and rehydration. Appl Environ Microbiol 71(12):8752–8763, https://doi.org/10.1128/AEM.71.12.8752-8763.2005

Soberón J, Nakamura M (2009) Niches and distributional areas: Concepts, methods, and assumptions. Proc Natl Acad Sci USA 106(Supplement 2):19,644–19,650, https://doi.org/10.1073/pnas.0901637106

Suchanek FM, Kasneci G, Weikum G (2007) YAGO: A core of semantic knowledge. In: Proceedings of the 16th International Conference on World Wide Web (WWW'07), pp 697–706, https://doi.org/10.1145/1242572.1242667

Tenenbaum JB, de Silva V, Langford JC (2000) A global geometric framework for nonlinear dimensionality reduction. Science 290(5500):2319–2323, https://doi.org/10.1126/science.290.5500.2319

Thuiller W, Lafourcade B, Engler R, Araújo MB (2009) BIOMOD – A platform for ensemble forecasting of species distributions. Ecography 32(3):369–373, https://doi.org/10.1111/j.1600-0587.2008.05742.x

Watts A, Ke D, Wang Q, Pillay A, Nicholson-Weller A, Lee JC (2005) Staphylococcus aureus strains that express serotype 5 or serotype 8 capsular polysaccharides differ in virulence. Infect Immun 73(6), https://doi.org/10.1128/IAI.73.6.3502-3511.2005

Wu H, Vreeken J, Tatti N, Ramakrishnan N (2014) Uncovering the plot: Detecting surprising coalitions of entities in multi-relational schemas. Data Min Knowl Disc 28(5–6):1398–1428, https://doi.org/10.1007/s10618-014-0370-1

Yan X, Han J (2002) gSpan: Graph-based substructure pattern mining. In: Proceedings of the 2002 IEEE International Conference on Data Mining (ICDM'02), pp 721–724, https://doi.org/10.1109/ICDM.2002.1184038

Zhao L, Zaki MJ, Ramakrishnan N (2006) BLOSOM: A framework for mining arbitrary Boolean expressions. In: Proceedings of the 12th ACM SIGKDD International Conference on Knowledge Discovery and Data Mining (KDD'06), pp 827–832, https://doi.org/10.1145/1150402.1150511

Zinchenko T, Galbrun E, Miettinen P (2015) Mining predictive redescriptions with trees. In: IEEE International Conference on Data Mining Workshops, pp 1672–1675, https://doi.org/10.1109/ICDMW.2015.123